Qualitative Analysis

the text of this book is printed
on 100% recycled paper

About the Author

Ray U. Brumblay received his A.B. degree from Indiana University, and his Ph.D. degree from the University of Wisconsin. For five years he taught at the Indiana University Extension Division in its Calumet Center at East Chicago, Indiana. After a period in the army, he joined the faculty of the University of Wisconsin—Milwaukee, where he is now Professor of Chemistry.

The author is a member of Sigma Xi and the American Chemical Society, and has contributed to *Analytical Chemistry*. He is also the author of *Quantitative Analysis,* in the College Outline Series.

COLLEGE OUTLINE SERIES

QUALITATIVE ANALYSIS

Ray U. Brumblay
Professor of Chemistry
University of Wisconsin — Milwaukee

BARNES & NOBLE BOOKS
A DIVISION OF HARPER & ROW, PUBLISHERS
New York, Evanston, San Francisco, London

L. C. catalogue card number: 64–19785

ISBN 389 00106 6

75 76 77 78 79 80 12 11 10 9 8 7 6 5

Manufactured in the United States of America

Preface

It is hoped that this book will provide the student with a brief, easily understood presentation of the principles and laboratory work usually contained in a course in qualitative analysis. Special attention has been given to problems which illustrate principles applied in the laboratory procedures. The step-wise solutions of problems are intended to make clear the reasoning behind the simple calculations necessary.

In order to make this book helpful to as many students as possible, those procedures and discussions of theory were included which provide another way of stating such material as is contained in most of the popular texts. In writing with this goal in mind it was thought desirable to simplify much of the language used. The author hopes that the simplifying process has resulted in discussions which will be easily understood but will not be lacking in rigor of thought.

Special mention should be made of Chapter 10, Mathematical Operations. The more complicated operations of mathematics used in this book are discussed briefly there, and they are illustrated with examples. If some mathematical step is not clear as given in the examples in other parts of the book, the student should consult Chapter 10 for a clarification of the operation applied. The use of logarithms is discussed at some length because of their application to many of the problems throughout the book.

Like most books, this one is the result of the work and help of many people. A great deal of gratitude is due to Dr. Gladys Walterhouse, editor, of Barnes & Noble, Inc., for her constant, valued assistance. The author is especially grateful to Mr. Wilbur L. Baker of Emory University for the many excellent suggestions he made and for the tremendously helpful work he did in locating errors in the manuscript.

Other books in the College Outline Series that can be of aid in studying the principles of Qualitative Analysis are:

Algebra
First-Year College Chemistry
Chemistry Problems and How to Solve Them
Quantitative Analysis
Physical Chemistry

Table of Contents

Qualitative Analysis

1

Introduction and Fundamental Facts

Qualitative analysis is one branch of analytical chemistry; the other branch is quantitative analysis. As the names indicate, qualitative analysis is concerned with identifying the constituents in a material and quantitative analysis is concerned with finding how much of one or more constituents is in a material.

Unlike many years ago, today qualitative analysis is studied by others than prospective industrial chemists, because it has been recognized as an aid to the understanding of many of the simpler principles of chemistry. In addition, the experience in using common apparatus and in determining the presence or absence of substances is excellent training and self-discipline.

The topics of chemistry well illustrated by qualitative analysis are:

1. The physical properties of substances, such as solubility and color.

2. The chemical properties of the common metals, nonmetals, and their compounds.

3. Chemical equilibrium, the study of opposed reactions at the point where the rates or speeds of the opposed reactions are equal and the reaction seems to have ceased at some point short of completion. Equilibria are encountered in the study of acids, bases, oxidation-reduction, and saturated solutions of slightly soluble salts. The last involves a special kind of equilibrium where the solubility product constant can be applied.

The General Plan of Qualitative Analysis

No haphazard trying of first one test, then another is effective in analyzing a substance qualitatively. An organized procedure must be followed. In detecting the presence or absence of metals or metal ions (cations), the following steps are taken:

1. All metals or salts are converted to salts that are soluble in water or in dilute nitric acid.

2. To a dilute nitric acid solution of these salts hydrochloric acid solution is added to precipitate out all cations whose chlorides are insoluble. These ions are called the Group I, or silver group, ions. The precipitated ions are further separated and tested for.

3. To the solution from Group I, sulfide is added to precipitate out all ions whose sulfides are insoluble in dilute acid. These ions are called the Group II, or copper-arsenic group, ions. The precipitate is treated so as to further separate the individual ions and to test for them.

4. The solution left from precipitating Group II is made alkaline with ammonia, and sulfide ions are added to precipitate out hydroxides and sulfides that are insoluble in basic solution. These ions are called the Group III, or aluminum-nickel group, ions. This precipitate is also further treated to separate and test for each ion.

5. Carbonate ions are added to the solution from the precipitation of Group III to precipitate out the carbonates of the Group IV, or barium group, ions. This precipitate is also further treated to separate and test for each ion in the group.

6. The solution remaining may contain only those ions, such as Na^+, K^+, Mg^{+2}, and NH_4^+, which form few insoluble salts. These are tested for individually.

The general plan for separating the cations into groups is shown in Chart 1.

Chart 1 indicates the operations for separating *all* of the common cations into groups. The only time a student or analyst is likely to have all of the ions together is when a "known" solution, one made up to contain all possible ions, is analyzed. Ordinarily only a few ions are in a sample and only a few ions are likely to occur in each group. However, the analyst proceeds as if all possible ions are present, and only a lack of a precipitate when an ion should form one is assurance of the ion's absence.

Example. To a solution colored green by one or more cations, HCl is added. If no precipitate of Group I chlorides occurs, the analyst knows that no Ag^+ or Hg_2^{+2} and only a trace of Pb^{+2} could be in the solution. If the addition of sulfide ions to the acid solution gives a black precipitate, the analyst must proceed with analyzing Group II as if all the possible ions in the group have been precipitated out. If no precipitates form when the subsequent groups are supposed to precipitate, then all ions in the green

CHART 1

General Plan for Separating Metal Ions (Cations) into Groups

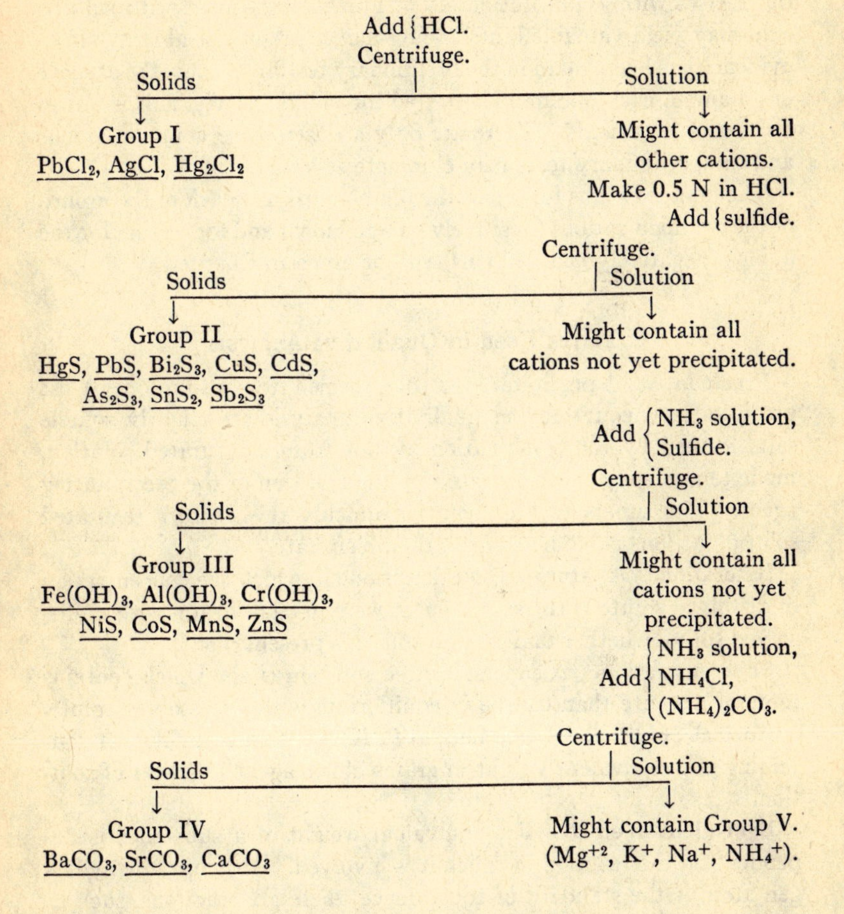

solution were in Group II and no procedures for the other groups are necessary.

The procedure for the analysis of substances for the anions, sometimes called "acid radicals" or "acid ions," is not quite so well organized as that for the cations. First, observations are made of certain properties, such as color, solubility in water, solubility in acids, and so on. Testing for solubility in acids may give valuable information. If the addition of hydrochloric acid solution to a bit of the sample does not cause bubbles of gas to be formed, then no sulfite or carbonate was present in the sample. Active metals and some sulfides produce gases with HCl solution also. Further treatment with sulfuric acid may yield valuable hints about what is present or absent. If no evidence has been found in the preliminary testing to show the absence of an anion, each one must be tested for individually. This is not so arduous as it sounds. There are only a dozen or so common anions and the preliminary tests may eliminate several of them.

The detailed plans for separating and identifying each of the groups of the common cations (positively charged ions) and for testing for the anions (negatively charged ions) will be given in Chapters 3–9.

Terms Used in Qualitative Analysis

Precipitate. A precipitate is a solid formed from a solution. Most precipitates encountered in qualitative analysis are slightly soluble salts formed by the combination of ions from a saturated solution, made temporarily supersaturated by the addition of the precipitating agent. The supersaturated solution quickly reverts to a saturated solution as ions combine to form the precipitate.

Saturation. A saturated solution is one in which the concentration of dissolved solute is the same as if it were in equilibrium with undissolved solute whether undissolved solute is present or not.

Supersaturation. A supersaturated solution is one which contains more of a solute than can be in equilibrium with undissolved solute.

Normal Solution. A one normal (1 N) solution is a solution containing one equivalent weight in grams of a reagent in a liter of solution.

Equivalent Weight. The equivalent weight of a substance is dependent on the reaction in which it is involved. Although the hydrogen atom is the standard of equivalence, there are reactions, such as oxidation-reduction reactions, in which no hydrogen is involved.

Equivalent Weights of Acids and Bases. In HCl there is one hydrogen atom which can ionize to yield hydrogen ions. The amount of HCl in grams that will yield one atomic weight of hydrogen as ions is the equivalent weight of HCl, 36.46 g. In H_2SO_4 there are 2 hydrogen atoms per molecule or 2 atomic weights of hydrogen per formula weight. One-half of the formula weight of sulfuric acid is its equivalent weight. For bases such as NaOH and $Ca(OH)_2$ each hydroxide is chemically equivalent to one hydrogen ion or atom. Therefore one formula weight of NaOH is the equivalent weight, and one-half the formula weight of $Ca(OH)_2$ is the equivalent weight.

Equivalent Weights of Salts. For finding the equivalent weight of a salt, divide the formula weight by either the total positive or the total negative valence.

Example 1. Na_2SO_4. The formula weight is about 142. The total of positive valence is +2 (2 sodium ions, each with a +1 charge). The total negative valence is −2 (the one sulfate ion is −2). Therefore divide 142 by 2 to find the equivalent weight of Na_2SO_4, which is 71. 71 g of Na_2SO_4 dissolved in water and diluted to 1 liter will produce a 1 N solution. 7.1 g of sodium sulfate in a liter of solution produces a 0.1 N (one-tenth normal) solution.

Example 2. $Al_2(SO_4)_3$. The total of positive charges is 2 × +3, or 6, since there are 2 Al^{+3} ions. The 3 sulfate (SO_4^{-2}) ions give a total of 3 × −2, or −6. Thus the equivalent weight is the formula weight divided by 6:

$$\frac{Al_2(SO_4)_3}{6} = \frac{342}{6} 57.0 \text{ g},$$

the equivalent weight of aluminum sulfate.

Equivalent weights and normality of oxidizing and reducing agents are discussed on p. 10.

Molar Solution. A one molar (1 M) solution is a solution containing one formula weight in grams of a substance in 1 liter of solution. The substance may be a molecular compound (not ionized), an ionic compound, or an ionic species, so that a special notation must be used to express molar concentrations of substances. The notation is enclosure of the formula in brackets.

Examples:

$[HCl]$ is read "molar concentration of hydrochloric acid."

$[Na_2SO_4]$ is read "molar concentration of sodium sulfate."

$[H_3O^+]$ is read "molar concentration of hydronium ion."

$[S^{-2}]$ is read "molar concentration of sulfide ion."

The concentrations may not always be one molar. For example:

$[HCl] = 0.1$ is read "the hydrochloric acid concentration equals (*or* is) one-tenth molar."

$[H_3O^+] = 0.001$ (*or* 1×10^{-3}) is read "the concentration of hydronium ions equals (*or* is) one-thousandth molar, or one times ten to the minus three molar."

A solution which is 1 molar in HCl contains one formula weight of HCl in each liter of solution or 36.5 g per liter. A 0.1 M solution of HCl contains 3.65 g per liter.

A 1.0 M solution of Na_2SO_4 contains one formula weight (142 g) of Na_2SO_4 per liter of solution. A 0.1 M solution would contain 14.2 g per liter and would be 0.1 M in sulfate ions but 0.2 M in sodium ions. There are 2 sodium ions for every sulfate ion dissolved.

Formal Solution. A one formal (1 F) solution is a solution made by dissolving one formula weight in grams of a substance in a solvent and diluting to 1 liter. The term *formal* is relatively new; it can be applied to solutions of all compounds, ionic or otherwise, and is the proper term for expressing the concentrations of substances such as salts. The term *molar* should be applied to designate the concentration of molecules in solution and should not be applied to solutions of substances which are largely in the form of ions in solution. However, usage has established that concentrations of ions and ionic substances, as well as molecular compounds, will be designated in terms of molarity unless some specific reason requires the use of the *formal* designation.

Oxidation and Reduction. *Oxidation* and *reduction* are terms which apply to changes in valence or oxidation number during a reaction. If an element loses electrons it acquires a higher, or more positive, oxidation number (or valence) and is said to be "oxidized." If an element gains electrons it acquires a lower, or more negative, oxidation number (or valence) and is said to be "reduced."

Example. In the analysis for the Group II metals, several sulfides, such as PbS, CuS, Bi_2S_3, and others, are dissolved with nitric acid, leaving HgS undissolved. The nitric acid (HNO_3) acts as an oxidizing agent as well as an acid in these reactions. As an acid it furnishes H_3O^+, and as an oxidizing agent if furnishes nitrate ions. HgS is so insoluble in water that nitric acid will not attack it. The sulfide of a metal, if it is to be attacked by HNO_3, must dissolve in water at least sufficiently to furnish the few sulfide ions required for nitrate to oxidize them. One possible reaction illustrating oxidation of sulfide by nitric acid is given by the equation:

$$3S^{-2} + 2NO_3^- + 8H_3O^+ \rightarrow 3S + 2NO\uparrow + 12H_2O.$$

The sulfide ion, S^{-2}, is oxidized from a -2 to a zero oxidation number, and the nitrogen in the nitrate, NO_3^-, is reduced from a $+5$ to a $+2$ oxidation number. The metal ion or ions enter solution from the solid state — a physical, not a chemical, change — and do not appear in the equation. They can be included thus:

$$3PbS + 2NO_3^- + 8H_3O^+ \rightarrow 3Pb^{+2} + 2NO\uparrow + 3S + 12H_2O.$$

The Dilution Equation. It is often necessary to prepare a solution of one concentration from another more concentrated one. Such a problem, and others concerning concentrations, can be solved by what is known as the "dilution" equation. This equation is an application of the following definition of concentration:

$$\text{Concentration} = \frac{\text{Weight of solute}}{\text{Volume of solution}}. \qquad (1\text{–}1)$$

Solving this for weight of solute:

$$\text{Weight of solute} = \text{Concentration} \times \text{Volume}.$$

If we want to know what volume of one solution of a known concentration must be used to prepare another volume with a lesser concentration, we reason that both solutions will have the same weight of solute in them. This is obvious, since if one solution is merely diluted with water to make the other, the total amount of dissolved solute will not change. If the weights of solute are equal in both solutions then:

$$\text{Concentration}_1 \times \text{Volume}_1 = \text{Weight of solute} = \text{Concentration}_2 \times \text{Volume}_2,$$

or more simply:

$$\text{Concentration}_1 \times \text{Volume}_1 = \text{Concentration}_2 \times \text{Volume}_2, \qquad (1\text{–}2)$$

where Concentration_1 and Volume_1 are for the first solution and Concentration_2 and Volume_2 are for the second solution. The concentration can be in any units of weight and volume, but the units must be the same for both solutions. The most common units are normalities or molarities, but grams per liter or pounds per gallon are satisfactory units also.

Example. To find the volume of concentrated HCl solution (12 N) which must be diluted to 50 ml to make 50 ml of 0.30 N solution, we merely substitute the values available in Equation (1–2) and solve for the volume of solution which is 12 N.

$$V \text{ ml} \times 12 \text{ N} = 50 \text{ ml} \times 0.30 \text{ N}.$$

$$V \text{ ml} = \frac{50 \times 0.30}{12} = 1.25 \text{ ml of } 12 \text{ N HCl}$$

must be diluted to 50 ml to produce a 0.30 N solution.

Instead of normality, the molarity may be used as the concentration of both solutions.

Example. What volume of 3 M sulfuric acid can be made up by diluting 5.0 ml of concentrated H_2SO_4 (18 M)? Applying Equation (1-2):

$$5.0 \text{ ml} \times 18 \text{ M} = V \text{ ml} \times 3 \text{ M}.$$

$$V \text{ ml} = \frac{5.0 \times 18}{3} = 30 \text{ ml of 3 M sulfuric acid}$$

can be prepared from 5.0 ml of 18 M H_2SO_4.

Balancing Equations. Chemical equations are shorthand statements which tell what substances react, what substances are produced, and the relative weights of reactants and products. In order that the relative weights may be correct, the equations must be balanced.

SIMPLE EQUATIONS. The simpler equations are balanced by inspection or, in a few cases, by applying the least common multiple principle.

Example 1. Many equations are simple combinations of ions in the ratio of one for one, thus:

$$Cd^{+2} + S^{-2} \rightarrow \underline{CdS}.$$
$$H_3O^+ + CO_3^{-2} \rightarrow HCO_3^- + H_2O.$$

The formulas need no coefficients (multipliers) for balancing.

Example 2. Many equations must be balanced by using only one coefficient:

$$Al^{+3} + 3OH^- \rightarrow \underline{Al(OH)_3}.$$
$$2H_3O^+ + S^{-2} \rightarrow H_2S\uparrow + 2H_2O.$$

Example 3. A fairly large number of equations must be balanced using coefficients of two or more of the formulas in the equation:

$$2Bi^{+3} + 3S^{-2} \rightarrow \underline{Bi_2S_3}.$$

One may apply the least common multiple principle to balance this equation. The charge on each ion is used as the coefficient of the other ion, giving a total of 6 as the sum of both positive and negative charges. With a little experience such a calculation can be performed at a glance.

REDOX REACTION EQUATIONS. A few equations for redox reactions can be balanced by inspection, but the more complicated ones require a special procedure. One effective method is the oxidation number method. The steps in this procedure are:

1. First write the unbalanced equation. For example:

$$S^{-2} + NO_3^- + H_3O^+ \rightarrow \underline{S} + NO\uparrow + H_2O.$$

This equation tells what things react and what things are produced, but because it is unbalanced it does not tell the relative weights of each.

2. The second step is to find the elements that have changed in oxidation state (valence) and what change has occurred for each element. In the example the S^{-2} has been oxidized to S, a change of 2 units of oxidation state or a loss of 2 electrons per atom. The nitrogen in the nitrate ion has been reduced from an oxidation state of +5 to one of +2, a gain of 3 electrons per atom. The other elements remain unchanged.

3. The third step is to balance these changes in oxidation state. This is most easily done by applying the least common multiple principle. The change in oxidation state of each element is used as a coefficient of the other. Thus:

The sulfide ion changed 2 units of oxidation. Use 2 as the coefficient or multiplier of the nitrogen. The nitrogen changed 3 units. Use 3 as the coefficient of the sulfur, thus:

$$3S^{-2} + 2NO_3^- + H_3O^+ \rightarrow \underline{3S} + 2NO\uparrow + H_2O.$$

The changes in oxidation state are now balanced. Very often the remaining unbalanced electrical charges, the hydrogen and the oxygen, can be balanced by inspection. If not, the next steps are necessary.

4. The fourth step is to balance the electrical charges on ions so that there are the same number of the same sign on each side of the equation:

On the left side of the equation there is a total of 8 negative charges, 2 from the nitrates and 6 from the 3 sulfides. There are no charged ions on the right; therefore the 8 negatives on the left must be equalized, either by adding 8 negative charges on the right in the form of some ion such as hydroxide or by adding 8 positive charges on the left in the form of H_3O^+. Multiplying the H_3O^+ by 8 balances the charges and the equation becomes:

$$3S^{-2} + 2NO_3^- + 8H_3O^+ \rightarrow \underline{3S} + 2NO\uparrow + H_2O.$$

5. By inspection it is obvious that the 24 hydrogen atoms on the left will supply the hydrogen for 12 molecules of water, and the equation becomes:

$$3S^{-2} + 2NO_3^- + 8H_3O^+ \rightarrow \underline{3S} + 2NO\uparrow + 12H_2O. \qquad (1\text{–}3)$$

The equation should now be balanced. The next step is to check to see if it is.

6. Total the atoms of oxygen on both sides of the equation. The equal number (14) on both sides indicates that the equation is balanced. A recheck of the balancing of all of the atoms in an equation is advisable.

Normality and Oxidation. From the definition of normality given on p. 4, a one normal solution of nitric acid, HNO_3, contains one formula weight, about 63 g, dissolved and diluted to 1 liter. As an acid, HNO_3 is able to produce one hydronium ion per formula weight, and the equivalent weight is the same as the formula weight. As an oxidizing agent in the reaction stated by Equation (1–3), the nitrogen in the nitrate ion undergoes a change of 3 units of oxidation, equivalent to changing 3 hydrogen ions to hydrogen atoms. Each nitrate in this reaction is equivalent to 3 hydrogen atoms, and the equivalent weight of HNO_3 as an oxidizing agent, where NO is the product, is one-third the formula weight.

$$\frac{HNO_3}{3} = \frac{63\text{ g}}{3} = 21\text{ g},$$

the equivalent weight of nitric acid as an oxidizing agent in this case.

In another case concentrated nitric acid reacts with copper to produce NO_2 instead of NO, thus:

$$Cu + 2NO_3^- + 4H_3O^+ \rightarrow Cu^{+2} + 2NO_2\uparrow + 6H_2O. \qquad (1\text{–}4)$$

The nitrogen in nitric acid here has been reduced in oxidation state from +5 in the nitrate to +4 in the NO_2, a change of 1 unit of oxidation state. One unit of oxidation state is equivalent to 1 atomic weight of hydrogen ion being reduced to hydrogen. The formula weight of nitric acid in Equation (1–4) is therefore the same as the equivalent weight.

The equivalent weight, based on oxidation-reduction reactions, is therefore dependent on the amount of change in oxidation number of the element in the reagent that reacts. The change in oxidation state, in turn, depends on the conditions such as temperature, concentration of the reactants, and the nature of the reactants themselves. Nitric acid in more dilute solutions reacts with zinc to yield ammonia, in which the oxidation state of nitrogen is −3. This is a change of 8 units of oxidation, from +5 to −3. For this change the equivalent weight of HNO_3 is one-eighth of the formula weight:

$$\frac{HNO_3}{8} = \frac{63}{8} = 7.875\text{ g},$$

the equivalent weight of nitric acid.

With zinc and more concentrated solutions of nitric acid, NO is the main product exactly as in Equation (1–3), p. 9. The equivalent weight of nitric acid then is one-third its formula weight.

Types of Redox Reactions in Qualitative Analysis

It seems advisable to summarize and classify some of the types of reactions in which oxidation states change during the qualitative analysis procedure.

1. In order to prepare alloys for analysis they are "dissolved" by converting the metals in them to soluble salts by the action of acids. The active metals, those more active than hydrogen, react with non-oxidizing acids such as HCl or dilute H_2SO_4 to yield hydrogen. Zinc is typical of such metals and the reaction can be written thus:

$$Zn + 2H_3O^+ \rightarrow Zn^{+2} + H_2\uparrow + 2H_2O.$$

The zinc is oxidized by the hydronium ion. Sn, Cd, Fe, Al, Mg, and other active metals react similarly.

Oxidizing acids such as HNO_3 and hot concentrated H_2SO_4 will react with the active metals and with many metals below hydrogen in activity as well. Instead of hydrogen being produced, the negative ion of the acid is reduced, nitric acid yielding NO if the acid is not very concentrated and NO_2 if it is concentrated. Sulfuric acid produces SO_2 when it acts as an oxidizing agent. Some examples are:

a. Copper with dilute HNO_3:

$$3Cu + 2NO_3^- + 8H_3O^+ \rightarrow 3Cu^{+2} + 2NO\uparrow + 12H_2O.$$

b. Copper with concentrated HNO_3:

$$Cu + 4H_3O^+ + 2NO_3^- \rightarrow Cu^{+2} + 2NO_2\uparrow + 6H_2O.$$

c. Copper with hot concentrated H_2SO_4:

$$Cu + 4H_3O^+ + SO_4^{-2} \rightarrow Cu^{+2} + SO_2\uparrow + 6H_2O.$$

Zn, Hg, Pb, Fe, and many other metals react similarly. Tin and antimony react differently, with HNO_3. Instead of being changed to soluble salts they are converted to very insoluble hydrated oxides, often called acids:

$$3Sn + 4H_3O^+ + 4NO_3^- \rightarrow \underline{3SnO_2(H_2O)} + 4NO\uparrow + 3H_2O,$$

and for antimony:

$$6Sb + 10H_3O^+ + 10NO_3^- \rightarrow \underline{3Sb_2O_5 \cdot (H_2O)} + 10NO\uparrow + 12H_2O.$$

See p. 122 for equations with other possible products. The insoluble acids or oxides are filtered off and are separated from the other metals that are left in solution so that in the analysis of an alloy, Sn and Sb do not occur in the regular procedure.

2. In the test for the mercurous ion in the Group I analysis, the mercury is both oxidized and reduced by what is sometimes called "auto-oxidation and reduction" or "disproportionation":

$$Hg_2Cl_2 + 2NH_3 \rightarrow \underline{Hg(NH_2)Cl} + \underline{Hg} + NH_4^+ + Cl^-.$$

In Hg_2Cl_2, mercury is in the +1 oxidation state. In $Hg(NH_2)Cl$, the mercury is in the +2 state, and in Hg it is in the zero oxidation state.

3. In Group II all of the common sulfides except HgS are dissolved in nitric acid. HgS is so insoluble in water or in acid solutions that the sulfide is not attacked by nitric acid. The other sulfides are dissolved away from the HgS with nitric acid, thus separating mercury from the other metals of the copper group. The reaction and balancing of the equation are discussed on pp. 6 and 8.

4. In the test for antimony in the Group II analysis metallic tin reduces antimony ions to the metallic state. This type of redox reaction is known as *displacement;* the tin displaces the antimony from the solution:

$$2Sb^{+3} + 3Sn \rightarrow \underline{2Sb} + 3Sn^{+2}.$$

5. Sulfide ions can be oxidized by metal ions as in Group III, which accounts for the cloudy solutions and whitish, yellowish, and grayish precipitates encountered there:

$$2Fe^{+3} + S^{-2} \rightarrow 2Fe^{+2} + \underline{S}.$$

6. Peroxide ions from hydrogen peroxide are used to oxidize chromic ions to chromate in Group III. This reaction is applied in the separation of chromium from nickel, cobalt, and manganese where the solution is made sufficiently alkaline so that hydroxide ions are plentiful:

$$2Cr^{+3} + 3O_2^{-2} + 4OH^- \rightarrow 2CrO_4^{-2} + 2H_2O.$$

Other redox reactions are applied in qualitative analysis, but these will be discussed more fully later.

Review Questions and Problems

1. Distinguish between qualitative and quantitative analysis.
2. What general principles of chemistry are illustrated by qualitative analysis operations?

3. What is the general plan for separating the ions in a solution into groups?
4. Using Chart I, in which group would each of the following ions be separated?
   a. Al^{+3}
   b. NH_4^+
   c. Hg_2^{+2}
   d. Hg^{+2}
   e. Pb^{+2}
   f. Co^{+2}
   g. Ca^{+2}
   h. K^+
5. Define:
   a. Precipitate.
   b. A one normal solution.
   c. A one molar solution.
   d. A 0.2 N solution.
   e. A 3 M solution.
   f. Oxidation.
   g. Oxidizing agent.
6. Calculate the equivalent weight of HCl, HNO_3, and H_2SO_4 as acids, and HNO_3 as an oxidizing agent where NO is the main product containing nitrogen.
7. Copper reacts with hot H_2SO_4 to produce SO_2 and Cu^{+2}. Write the balanced equation for the reaction and calculate the equivalent weight of H_2SO_4 as an oxidizing agent in this reaction.
8. $KMnO_4$ solution will oxidize ferrous ions (Fe^{+2}) to ferric ions (Fe^{+3}). At the same time the manganese is reduced to manganous ions (Mn^{+2}). Write the balanced equation and calculate the equivalent weight of $KMnO_4$ for this reaction.
9. What volume of 0.1091 N HCl solution will be needed to make up 250 ml of 0.1000 N HCl?
10. What is the normality of an HNO_3 solution if 25.00 ml of it are just neutralized by 31.15 ml of 0.0840 N NaOH solution?
11. Balance the following equations for redox reactions:
    a. $SO_2 + MnO_4^- + H_3O^+ \rightarrow Mn^{+2} + H_2SO_4 + H_2O$.
    b. $MnO_2 + S^{-2} + H_3O^+ \rightarrow SO_2\uparrow + Mn^{+2} + H_2O$.
    c. $NO_3^- + S^{-2} + H_3O^+ \rightarrow NO\uparrow + SO_2\uparrow + H_2O$.
    d. $Cr^{+3} + O_2^{-2} + OH^- \rightarrow CrO_4^{-2} + H_2O$.
    e. $Hg_2Cl_2 + NH_3 \rightarrow \underline{HgNH_2Cl} + NH_4^+ + Hg + Cl^-$.
    f. $HgCl_2 + Sn^{+2} + Cl^- \rightarrow \underline{Hg_2Cl_2} + SnCl_6^{-2}$.

2

Equilibrium in Qualitative Analysis

Many qualitative analysis procedures are concerned with reactions at equilibrium. In the general plan for precipitating the copper group ions as sulfides, the sulfide ion concentration is controlled by applying equilibrium principles to the interaction of sulfide ions, hydronium ions, and hydrogen sulfide molecules. Simultaneously the sulfide is also involved in an equilibrium with the ions of the metals and the solid sulfides that have precipitated. Wherever any one or more precipitates are formed in separating one element from another, some equilibrium situations are created to accomplish the desired result. In order to separate ions from each other, some are induced to form insoluble compounds while others remain in solution. The more insoluble the compounds are that precipitate, the more nearly complete the separation. Therefore we must study the factors that affect equilibria influencing the solubility of slightly soluble substances.

Factors Affecting the Solubility of Precipitates

In order to be able to create conditions under which precipitation reactions occur as completely as possible, the chemist must understand the principles of equilibrium reactions. Precipitation reactions are never quite complete because every precipitate is at least minutely soluble. The concern of the chemist is to control conditions near the point of completion.

The controllable conditions which affect solubilities of salts in saturated solutions are:

1. Relative molar concentrations of the ions of the salt in solution, the "common ion effect" (see p. 19).
2. Complex ion formation.
3. If anions of weak acids are part of the dissolved salt, the pH of the solution is important.

These three factors, as well as the temperature and activities, can be controlled by the analyst within certain limits. The nature of a salt is beyond control, although it is the most important factor of all affecting solubility.

The Law of Mass Action. The law of mass action is a special case of the principle of Le Chatelier, which states that when a stress is applied to a system at equilibrium the equilibrium shifts (is displaced) in such a way as to reduce or relieve the stress applied. The law of mass action is concerned only with stresses applied to equilibria by changes in mass or quantity of the reactants. Such masses are expressed in terms of molar concentration, that is, moles per liter of the reactants in solutions. An equilibrium is a situation in which two reactions opposing each other are occurring simultaneously and the two opposing rates are equal:

$$\text{Rate} \rightarrow = \text{Rate} \leftarrow.$$

The law of mass action states that the rates of reactions are proportional to the products of the active concentrations of the reactants, each concentration being taken to a power equal to its coefficient in the balanced equation for the reaction. Thus for the general reaction at equilibrium:

$$aA + bB \cdots \rightleftharpoons cC + dD \cdots$$

$$\text{Rate toward right:} \quad R_1 \propto [A]^a \times [B]^b \cdots$$

$$\text{Rate toward left:} \quad R_2 \propto [C]^c \times [D]^d \cdots$$

Brackets mean gram moles per liter, and the symbol $\propto$ means "is proportional to."

To change the proportions to equations, rate constants k_1 and k_2 are introduced. These constants represent the effects of the nature of the reactants, catalysts, temperature, pressure, and all other factors — except concentrations — that affect speeds of reactions. Then:

$$R_1 = k_1[A]^a[B]^b \cdots$$
$$R_2 = k_2[C]^c[D]^d \cdots$$

Since R_1 and R_2 are equal at equilibrium and things equal to the same things are equal to each other, we can write:

$$k_1[A]^a[B]^b \cdots = k_2[C]^c[D]^d \ldots$$

Collecting all constants on one side and all variables on the other:

$$\frac{k_1}{k_2} = \frac{[C]^c[D]^d \cdots}{[A]\,[B]^b \cdots}$$

A ratio of two constants is a constant, so the ratio k_1/k_2 is replaced by K_e, giving

$$K_e = \frac{[C]^c[D]^d \cdots}{[A]^a[B]^b \cdots} \tag{2-1}$$

This is the general form of the equilibrium constant. It can be applied to many kinds of reactions such as those of ionization, decomposition, and precipitation. The three dots imply that if there are more than two reactants they should be included in the equation just as the others, and if there are less than two, only the one reactant would be indicated.

Solubility Products. An important application of the law of mass action is in precipitation reaction calculations.

The precipitation of silver chloride in Group I involves the equilibrium between solid silver chloride and silver and chloride ions:

$$AgCl(Solid) \rightleftharpoons Ag^+ + Cl^-.$$

The equilibrium constant for this reaction is:

$$K_e = \frac{[Ag^+][Cl^-]}{[AgCl](Solid)},$$

but at equilibrium the total quantity of solid present is immaterial because the solubility of the salt controls the concentration of the ions completely, regardless of the amount of solid. Because the solid is in the "standard state," the value for the concentration of AgCl solid can be taken as unity, leaving:

$$K_{sp} = [Ag^+][Cl^-], \tag{2-2}$$

where K_{sp} is the solubility product constant, or ion product constant, for silver chloride.

For a reaction such as As_2S_3 in equilibrium with its ions in a saturated solution, the coefficients of the ions are used as exponents exactly as in the equilibrium constant equation (2–1); thus for the equilibrium $As_2S_3 \rightleftharpoons 2As^{+3} + 3S^{-2}$:

$$K_{sp} = [As^{+3}]^2[S^{-2}]^3. \tag{2-3}$$

Calculating Solubility Products from Solubilities. The numerical values for solubility products are obtained from measurements of the solubility of the salts in water. Solubility values for salts in various solutions can be applied, but since exactly the same methods would be used, solubilities in pure water, except in special cases, will be considered here.

Example 1. The solubility of AgCl is 0.0016 g per liter at 20° C. Calculate the value of the solubility product constant for AgCl.

In a saturated solution, every mole of AgCl dissolved yields 1 mole each of Ag^+ and Cl^-, thus:

$$AgCl \rightleftharpoons Ag^+ + Cl^-$$

and the molar concentration of AgCl, [AgCl], is equal to the molar concentration of Cl^-, $[Cl^-]$, and to the molar concentration of Ag^+, $[Ag^+]$. The molar concentration of AgCl is found by dividing the concentration in grams per liter by the formula weight:

$$[AgCl] = \frac{0.0016}{143.34} = 0.000011 = 1.1 \times 10^{-5}.$$

Substituting in the formula for the K_{sp}, Equation (2–2):

$$\begin{aligned} K_{sp} &= [Ag^+][Cl^-] \\ &= 1.1 \times 10^{-5} \times 1.1 \times 10^{-5} \\ &= 1.2 \times 10^{-10}. \end{aligned}$$

Values for solubility product constants of various substances are given in Appendix VI.

Example 2. In a liter of a saturated solution of Ag_2CrO_4, 0.0322 g of silver chromate is in solution at 25° C. Calculate the value of its solubility product constant. The equilibrium is:

$$Ag_2CrO_4 \rightleftharpoons 2Ag^+ + CrO_4^{-2}.$$

The molar solubility is found by dividing the solubility in grams per liter by the formula weight:

$$[Ag_2CrO_4] = \frac{0.0322}{331.77} = 9.706 \times 10^{-5}.$$

For each mole of silver chromate dissolved, 2 silver ions and 1 chromate ion are produced. Therefore:

$$\begin{aligned} [CrO_4^{-2}] &= 9.706 \times 10^{-5}. \\ [Ag^+] &= 2 \times 9.706 \times 10^{-5} = 19.412 \times 10^{-5}. \end{aligned}$$

Then, substituting in the K_{sp} thus:

$$\begin{aligned} K_{sp} &= [Ag^+]^2[CrO_4^{-2}] \\ &= (19.412 \times 10^{-5})^2(9.706 \times 10^{-5}) \\ &= 3.65 \times 10^{-12}. \end{aligned}$$

CALCULATING SOLUBILITIES FROM SOLUBILITY PRODUCTS. The solubility of a slightly soluble salt can be calculated from the solubility product.

Example 1. The K_{sp} of AgBr is 7.7×10^{-13}. Calculate its solubility in grams per 100 ml.

Since the molar concentrations of Ag^+ and Br^- are equal and each is equal to the molar solubility of AgBr, then:

$$[Ag^+] = [Br^-] = [AgBr] = \sqrt{7.7 \times 10^{-13}}$$
$$= 8.8 \times 10^{-7} \text{ moles per liter,}$$

the concentration of AgBr in a saturated solution. To change moles to grams, multiply by the formula weight of AgBr, 188:

$$8.8 \times 10^{-7} \times 188 = 0.000165 \text{ g per liter.}$$

Since 100 ml is 0.1 as large as a liter:

$$1.65 \times 10^{-4} \times 0.1 = 1.65 \times 10^{-5} \text{ g per 100 ml.}$$

Example 2. Calculate the solubility in grams per 100 ml of $PbCl_2$. Its solubility product constant is 1×10^{-4}, and the equilibrium equation is:

$$PbCl_2 \rightleftharpoons Pb^{+2} + 2Cl^-.$$

The molar solubility of lead chloride will be the same as the molar concentration of lead ions, $[Pb^{+2}]$, and will be one-half the molar concentration of chloride ions, $[Cl^-]$, because for every mole of $PbCl_2$ dissolved, $1Pb^{+2}$ and $2Cl^-$ are formed. Therefore, if M = molar solubility of $PbCl_2$:

$$2M = [Cl^-].$$
$$M = [Pb^{+2}].$$

The solubility product constant for lead chloride is:

$$[Pb^{+2}][Cl^-]^2 = 1 \times 10^{-4}.$$

Then:

$$M(2M)^2 = 1 \times 10^{-4}.$$
$$4M^3 = 1 \times 10^{-4}.$$
$$M^3 = 0.25 \times 10^{-4}.$$
$$M = \sqrt[3]{25 \times 10^{-6}}.$$
$$\log M = \frac{\log (25 \times 10^{-6})}{3}.$$
$$M = 2.92 \times 10^{-2}$$
$$= 0.0292 \text{ mole of lead chloride per liter in a saturated solution.}$$

To find the solubility in grams, multiply the number of moles by the weight in grams of each mole:

$$0.0292 \times 278.12 = 8.12 \text{ g, the solubility of lead chloride per liter.}$$

Because 100 ml is 0.1 liter, 8.12 g $\times$ 0.1, or 0.812 g, is the solubility of $PbCl_2$ in grams per 100 ml.

Example 3. Using 1.1×10^{-33} as the K_{sp} of As_2S_3, calculate its solubility in grams per liter.

As in Example 2, let M = molar solubility of As_2S_3.

Then:

$$2M = [As^{+3}].$$
$$3M = [S^{-2}].$$
$$(2M)^2 \times (3M)^3 = 1.1 \times 10^{-33}.$$
$$108M^5 = 1.1 \times 10^{-33}.$$
$$M^5 = \frac{1.1 \times 10^{-33}}{108}.$$
$$M^5 = 1.018 \times 10^{-35}.$$
$$\log M = \frac{\log (1.018 \times 10^{-35})}{5}.$$
$$M = 1.004 \times 10^{-7} \text{ moles per liter of } As_2S_3 \text{ in a saturated solution.}$$

To convert moles per liter to grams per liter, multiply the molar concentration by the formula weight:

$$1.004 \times 10^{-7} \times 246.02 = 247 \times 10^{-7}$$
$$= 2.47 \times 10^{-5} \text{ g per liter of } As_2S_3 \text{ in a saturated solution.}$$

THE COMMON ION EFFECT. Each ion in a salt is said to be "common" to that salt. In $PbCl_2$ both Pb^{+2} and Cl^- are ions common to $PbCl_2$, and in Ag_2CrO_4 both Ag^+ and CrO_4^{-2} are common ions of Ag_2CrO_4. In saturated solutions of salts an increase in the concentration of one ion causes a decrease in the concentration of the other and vice versa.

One can calculate the effect that a small excess of one of the ions of a slightly soluble salt will have on the concentration of the other ion in a saturated solution of the salt.

Example 1. The K_{sp} of Ag_2CrO_4 is 3.65×10^{-12}. Calculate the molar concentration of Ag^+ in a water solution saturated with Ag_2CrO_4 and 0.02 molar in chromate ion.

The solution of this type of problem requires a simple substitution in the formula for the solubility product constant. The equation for the K_{sp} of Ag_2CrO_4 is:

$$[Ag^+]^2[CrO_4^{-2}] = 3.65 \times 10^{-12}.$$

Substituting 0.02 for the chromate concentration:

$$[Ag^+]^2 \times 0.02 = 3.65 \times 10^{-12}.$$

$$[Ag^+]^2 = \frac{3.65 \times 10^{-12}}{0.02} = 182.5 \times 10^{-12}.$$

$$[Ag^+] = \sqrt{182.5 \times 10^{-12}}.$$

$$\log [Ag^+] = \frac{\log (182.5 \times 10^{-12})}{2}.$$

$$[Ag^+] = 13.5 \times 10^{-6} = 1.35 \times 10^{-5}.$$

Example 2. The silver ion concentration in a solution saturated with silver chromate is found as follows:

Let M = the molar concentration of the Ag_2CrO_4 in the saturated solution. Then:

$$2M = [Ag^+].$$

$$M = [CrO_4^{-2}].$$

$$(2M)^2M = 3.65 \times 10^{-12}.$$

$$4M^3 = 3.65 \times 10^{-12}.$$

$$M^3 = 0.91 \times 10^{-12}.$$

$$\log M = \frac{\log 910 \times 10^{-15}}{3}.$$

$$M = 9.69 \times 10^{-5}.$$

$$2M = [Ag^+] = 19.38 \times 10^{-4} = 1.94 \times 10^{-3}.$$

Thus, the small excess of chromate ion appreciably decreases the silver ion concentration.

Example 3. If the silver ion concentration is 0.02 molar in a solution saturated with silver chromate, what will the molar concentration of the CrO_4^{-2} be? Substituting in the solubility product constant formula:

$$[Ag^+]^2[CrO_4^{-2}] = 3.65 \times 10^{-12}.$$

$$[0.02]^2[CrO_4^{-2}] = 3.65 \times 10^{-12}.$$

$$[CrO_4^{-2}] = \frac{3.65 \times 10^{-12}}{0.0004}.$$

$$= 0.9125 \times 10^{-8}.$$

$$= 9.125 \times 10^{-9} = 9.1 \times 10^{-9}.$$

Solubility products and values derived from them are seldom accurate to more than two significant figures. Therefore the 9.125×10^{-9} can be rounded off to 9.1×10^{-9}. However, the third significant figure frequently is carried along as a doubtful figure.

Complex Ion Formation. In qualitative analysis advantage is often taken of the fact that some ions form complex ions that are stable and much more soluble than compounds formed by uncomplexed ions. Certain ions are thereby separated from others. The silver ion in Group I is separated from mercurous ions by forming the very soluble ammonia-silver complex, leaving the mercurous chloride behind as an insoluble residue.

THE NATURE OF COMPLEX IONS. Complex ions are compounds made up of a cation with other ions or molecules. The parts of the complex are partly held together by coordinate covalent bonds formed by the sharing of a pair of electrons between the central cation and the other ion or molecule. In no case does the cation furnish the pair of electrons shared; these are furnished by the ion or molecule other than the central cation. The cation is therefore a Lewis acid; the other ions or molecules are the Lewis base. (See pp. 27–28.) On the basis of the number of particles involved, the hydronium ion, H_3O^+, is probably the simplest complex ion. The hydrogen ion acts as the Lewis acid, and water as the Lewis base.

COMPLEXES WITH AMMONIA. Copper ions form a complex with ammonia which is deep blue in color and very soluble in water. In the copper group analysis (Step 13, p. 71), bismuth ions are separated from copper and cadmium ions by adding ammonia solution to a solution containing all three metals. The bismuth does not form a complex ion with ammonia, so it precipitates as the hydroxide. Copper and cadmium form soluble complexes with ammonia and remain in solution. A diagram of the copper ammonia complex is:

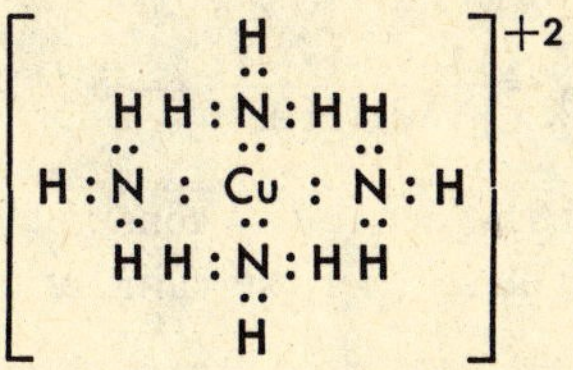

```
┌              H             ┐+2
               ..
         H H : N : H H
         ..    ..    ..
     H : N  :  Cu  :  N : H
         ..    ..    ..
         H H : N : H H
               ..
               H
└                            ┘
```

Cadmium, nickel, cobalt, zinc, and many other metal ions form similar complexes with varying numbers of ammonia molecules. Other molecules or ions such as water, cyanide, chloride, fluoride, and oxalate can replace ammonia if conditions are right. In fact, complexes are probably formed from one another in solution by replacement of one ion or molecule by another, in steps. Such steps are indicated by the following equations, in which water is replaced by ammonia in a series of equilibria, all of which are in operation at the same time, and each reaction is more nearly complete toward the right in the higher concentration of ammonia:

$$Cu(H_2O)_4^{+2} + NH_3 \rightleftharpoons Cu(H_2O)_3(NH_3)^{+2} + H_2O.$$
$$Cu(H_2O)_3(NH_3)^{+2} + NH_3 \rightleftharpoons Cu(H_2O)_2(NH_3)_2^{+2} + H_2O.$$
$$Cu(H_2O)_2(NH_3)_2^{+2} + NH_3 \rightleftharpoons Cu(H_2O)(NH_3)_3^{+2} + H_2O.$$
$$Cu(H_2O)(NH_3)_3^{+2} + NH_3 \rightleftharpoons Cu(NH_3)_4^{+2} + H_2O.$$

The ammonia complexes are less dissociated into ions and molecules than the water complexes and are said to be more "stable" than the corresponding water complexes. Cyanide complexes, where cyanide ions replace water molecules, are generally more stable than the corresponding ammonia complexes.

EQUILIBRIUM AND COMPLEX IONS. A study of the reactions by which complex ions are formed shows that such reactions are indeed equilibrium reactions. The law of mass action can be applied to them, and equilibrium constants can be determined. Such constants are termed *instability constants* because the equations for the reactions are generally written indicating decomposition rather than formation of the complex. The value for an instability constant is the reciprocal of the ordinary equilibrium constant. These decomposition equations are also usually written without showing water of hydration (complexes with water) thus:

$$Cu(NH_3)_4^{+2} \rightleftharpoons Cu^{+2} + 4NH_3.$$

The instability constant equation for this reaction, which is the sum of all of the steps in the equations shown on p. 21, is written thus:

$$K_{\text{Inst}} = \frac{[Cu^{+2}][NH_3]^4}{[Cu(NH_3)_4^{+2}]} = 5 \times 10^{-14}. \qquad (2\text{–}4)$$

In general, the size of the instability constant indicates the stability of the complex ion. The smaller the value of the constant, the more stable the complex ion.

A few other equilibrium equations showing formation and decomposition of complex ions are shown in Appendix III along with instability constants for them.

Instability constants are of value in calculating the concentrations of either the coordinating groups or the central ion. It is possible to show why sulfide precipitates cadmium sulfide but not copper sulfide from cyanide solutions. Cadmium and cupric copper ions react differently with cyanide:

$$Cd(NH_3)_4^{+2} + 4CN^- \rightarrow Cd(CN)_4^{-2} + 4NH_3.$$

This is a simple replacement of ammonia molecules by cyanide ions. In the case of copper, the most probable reaction in the alkaline solution is:

$$2Cu(NH_3)_4^{+2} + 7CN^- + 2OH^- \rightarrow 2Cu(CN)_3^{-2} + CNO^- + 8NH_3.$$

This is an oxidation-reduction reaction in which Cu^{+2} is reduced to Cu^+ by CN^- and CN^- is oxidized to cyanate, CNO^-. A reaction frequently reported to take place here is the oxidation of CN^- to cyanogen, $(CN)_2$, by Cu^{+2}. The equation is:

$$2Cu(NH_3)_4^{+2} + 8CN^- \rightarrow 2Cu(CN)_3^{-2} + (CN)_2 + 8NH_3.$$

Example 1. Calculate the concentration of Cu^+ in a solution that is 0.2 M in CN^- and 0.03 M in $Cu(CN)_3^{-2}$ and show that Cu_2S will not precipitate from such a solution if the solubility product constant for Cu_2S is 2×10^{-48}.

Step 1. Substitute the values given in the equation for the instability constant for $Cu(CN)_3^{-2}$. The equation for the instability constant is:

$$K_{\text{Inst}} = \frac{[Cu^+][CN^-]^3}{[Cu(CN)_3^{-2}]} = 5 \times 10^{-28}.$$

Substitution of values given in the problem gives:

$$\frac{[Cu^+](0.2)^3}{0.03} = 5 \times 10^{-28}.$$

Step 2. Solve for $[Cu^+]$, the molar concentration of cuprous copper:

$$[Cu^+] = \frac{5 \times 10^{-28} \times 3 \times 10^{-2}}{8 \times 10^{-3}} = 1.88 \times 10^{-27}.$$

If 1.88×10^{-27} is substituted in the solubility product constant equation for the concentration of cuprous ions, the concentration of sulfide needed to saturate the solution with Cu_2S can be found. The solubility product constant for Cu_2S is:

$$[Cu^+]^2[S^{-2}] = 2 \times 10^{-48}.$$

Step 3. Substitute the value found for the $[Cu^+]$ in the solubility product equation:

$$(1.88 \times 10^{-27})^2[S^{-2}] = 2 \times 10^{-48}.$$

Step 4. Solve for the sulfide ion concentration:

$$[S^{-2}] = \frac{2 \times 10^{-48}}{(1.88 \times 10^{-27})^2} = \frac{2 \times 10^{-48}}{3.53 \times 10^{-54}}$$
$$= 0.566 \times 10^6 = 5.66 \times 10^5,$$

over 500,000 moles per liter of sulfide needed to saturate the solution with Cu_2S in a cyanide solution. Such a huge concentration is ridiculously impossible. A saturated solution of H_2S is about 0.1 M, and concentrations of 5 or 10 molar sulfide are possible in strongly alkaline solutions. Therefore cuprous sulfide cannot precipitate from solutions containing cyanide.

Example 2. Calculate the molar concentration of cadmium ions in a solution that is 0.2 M in CN^- and 0.03 M in $Cd(CN)_4^{-2}$, and show that making such a solution 0.01 M in sulfide ions will precipitate CdS. The solubility product constant for CdS is 1×10^{-26}.

Step 1. Substitute such values as are given in the problem in the equation for the instability constant for $Cd(CN)_4^{-2}$. The equation is:

$$K_{\text{Inst}} = \frac{[Cd^{+2}][CN^-]^4}{[Cd(CN)_4^{-2}]} = 1.4 \times 10^{-17}.$$

Substitution gives:

$$\frac{[Cd^{+2}](0.2)^4}{0.03} = 1.4 \times 10^{-17}.$$

Step 2. Solve for $[Cd^{+2}]$:

$$[Cd^{+2}] = \frac{1.4 \times 10^{-17} \times 0.03}{0.0016} = 2.6 \times 10^{-16}.$$

Step 3. Substitute the value found for the cadmium ion concentration in the solubility product constant equation for CdS:

$$2.6 \times 10^{-16} \times [S^{-2}] = 1 \times 10^{-26}.$$

Step 4. Solve for the sulfide ion concentration that will saturate the solution with CdS:

$$[S^{-2}] = \frac{1 \times 10^{-26}}{2.6 \times 10^{-16}} = 0.385 \times 10^{-10} = 3.85 \times 10^{-11}.$$

This concentration, 3.85×10^{-11} moles per liter, is far less than the 0.01 M sulfide the problem states is available. The CdS, therefore, should precipitate from solutions containing cyanide.

COMPLEX ION NOMENCLATURE. The term *complex ion* is not correct in all cases. A better name for such compounds is *complex compounds* because not all are ions, some are neutral molecules, and all are to some degree the result of the formation of coordinate covalent bonds between some ions or molecules and a central ion. The coordinating groups, sometimes called "ligands," are those parts of a complex compound other than the central atom. They are the Lewis base (see p. 27) part of the complex because they furnish the pairs of electrons for the central atom to share, thus forming coordinate covalent bonds. A coordinate covalent bond is one where both the electrons of a shared pair are furnished by one of the atoms only.

A system has been developed for naming complexes. With the few rules given below, almost all of the common complex compounds can be named.

Rule 1. The coordinating groups are named first and in the order: negative ion, neutral molecule, positive ion.

Rule 2. The central element is named last with the ending *-ate* if the charge on the whole complex is negative and with the name of the element unchanged by any ending if the charge on the entire complex is positive or neutral.

Rule 3. The oxidation state or valence number of the central element is indicated by adding a Roman numeral in parentheses after the name of the element.

Rule 4. The prefixes *di-*, *tri-*, and *tetra-*, for "two," "three," and "four," precede the coordinating groups, the name of the coordinating group not being enclosed in parentheses if it is simple, such as *chloro-*, *ammine-*, or *nitro-*. If the coordinating group or ion is not simple, then the prefixes *bis-*, *tris-*, *tetrakis-*, etc., are preferred to *di-*, *tri-*, and *tetra-* and the name of the group is enclosed in parentheses.

Rule 5. The names of negative coordinating ions end in *-o*, such as *cyano* and *chloro*. The names of positive or neutral groups have no specific ending, but coordinated ammonia is called *ammine* and coordinated water is *aquo*.

Examples. The compound $Ag(NH_3)_2^+$ is diamminesilver(I).
$Cu(CN)_3^{-2}$ is tricyanocuprate(I).
$Cu(NH_3)_4^{+2}$ is the tetramminecopper(II) ion.
$Cd(CN)_4^{-2}$ is tetracyanocadmiate(II).
$Sn(OH)_4^{-2}$ is the tetrahydroxostannate(II) ion.
$SnCl_6^{-2}$ is the hexachlorostannate(IV) ion.
$Cu(NH_3)(H_2O)_2Cl^+$ is chloroamminediaquocopper(II).
The "ammine" is placed before "aquo" because of alphabetical listing rather than because of any rule given above, since both water and ammonia are neutral molecules. Where "en" represents ethylenediammine,
$Co(en)_2(Cl)(NO_2)^+$ is the chloronitrobis(ethylenediammine)cobalt(III) ion.

Frequently the rigidity of formal names is avoided by calling $Ag(NH_3)_2^+$ the *silver ammonia complex*, $Cu(CN)_3^{-2}$ the *cuprous cyanide complex*, etc. Such names are usually clear enough to avoid confusion and require no adherence to strict rules.

Complex Ion Structure. There is much yet to be learned about the structure of complex ions. But it is certain that electrostatic forces (attractions between oppositely charged bodies) help to hold the ligands around the central atom in a complex ion or molecule. It is just as certain that some covalency (sharing of electrons) also exists between the ligands and the central atom. In some cases the electrostatic or ionic forces seem to predominate, and in others some form of covalence seems to be of more importance in holding the ligands to the central ion.

The electrostatic forces occur (1) between ions of opposite charge, and (2) between ions and polar molecules. Polar molecules include such substances as water and ammonia, which have a negative pole and a positive pole. The negative pole (the negative end) of a molecule is attracted by positive ions and the positive pole by negative ions.

Other than electrostatic forces there are coordinate covalent bonds between ligands and the central ion. A coordinate bond begins to be formed when the ligand, with a pair of unshared electrons, approaches the central ion and begins to share its pair of electrons with the central ion. The two electrons from the ligand appear to enter regions around the central ion known as orbitals, and they oscillate between an orbital in the central ion and the orbital in the atom of the ligand where they originated.

ORBITALS. An orbital is a region in space around an atomic nucleus with a fairly definite shape. Each orbital may be occupied by 1 or by 2 electrons, but never more than 2, and when 2 electrons occupy the same orbital they are thought to have opposite spins.

Some orbitals, the first ones to be filled in each energy level, have no directional properties. These orbitals are spherical, and an ion or molecule attached to a central atom by sharing electrons in such an orbital would not be held in any fixed direction with regard to the central atom. All other orbitals—the second, third, and others in an energy level—have directional properties, causing ligands which unite with the central atom to do so at certain specific points about the central atom. This causes the resulting complex to have a geometrical shape in space, such as a tetrahedron, a plane, or an octahedron. The geometric configuration of complexes is interesting but not essential to the study of qualitative analysis. For a discussion of such structures, see L. E. Sutton, "Some Recent Developments in the Theory of Bonding in Complex Compounds of the Transition Metals," *Journal of Chemical Education*, XXXVII (1960), 498–505.

In qualitative analysis it is necessary to know how to apply complex ion formation in separating many metal ions from others. For examples of such applications, see pp. 23, 72, 74, 80, and 93.

Equilibrium in Acids and Bases

The study of neutralization reactions has resulted in several theories as to what acids and bases are and how they react. In addition, many laws have been discovered which can be expressed mathematically. The application of these laws to reactions in qualitative analysis is a large part of the theory of equilibrium in acids and bases.

Definitions

Two modern concepts express our ideas of acids and bases. The more general Lewis concept is seldom used in discussions of acid-base phenomena. The Brønsted and Lowry definitions are a modification of older concepts and are easily applicable to neutralization problems

The Lewis Electron-Pair Concept. The Lewis theory is generally regarded as the most widely applicable of the reasonable ideas about acids, bases, and neutralization. Its definitions are:

ACID. An acid is a substance which can share an unshared electron pair on an atom in another molecule or ion. The hydrogen ion, H^+, and the sulfur trioxide molecule, SO_3, can be considered acids since either can share a pair of electrons furnished by another atom.

BASE. A base is a substance which contains an atom having an electron pair which is capable of being shared by an atom in another molecule or ion. The hydroxide ion, OH^-, and the oxide ion, O^{-2}, are both bases because each has a pair of electrons available for sharing. The ammonia molecule is also a base.

NEUTRALIZATION. Neutralization is the formation of a coordinate covalent bond between an acid and a base, thus:

$$\begin{array}{ccccc} \text{Acid} & & \text{Base} & & \\ H^+ & + & :\underset{\cdot\cdot}{\ddot{O}}:H^- & \rightarrow & H:\underset{\cdot\cdot}{\ddot{O}}:H \end{array}$$

$$\begin{array}{ccccc} :\ddot{O}: & & & & :\ddot{O}: \\ \underset{\cdot\cdot}{\ddot{S}}:\underset{\cdot\cdot}{\ddot{O}}: & + & :\underset{\cdot\cdot}{\ddot{O}}:^{-2} & \rightarrow & :\underset{\cdot\cdot}{\ddot{O}}:\underset{\cdot\cdot}{\ddot{S}}:\underset{\cdot\cdot}{\ddot{O}}:^{-2} \\ :\underset{\cdot\cdot}{O}: & & & & :\underset{\cdot\cdot}{O}: \end{array}$$

$$\begin{array}{ccccc} & & H & & H \\ H^+ & + & :\underset{\cdot\cdot}{\ddot{N}}:H & \rightarrow & H:\underset{\cdot\cdot}{\ddot{N}}:H^+ \\ & & H & & H \end{array}$$

When a proton reacts with either a hydroxide ion or an ammonia molecule, a coordinate covalent bond is formed in making a water molecule or an ammonium ion. When an oxide ion unites with sulfur trioxide to form a sulfate ion, the oxide forms a coordinate covalent bond by sharing a pair of unshared electrons with the sulfur in the sulfur trioxide.

IMPORTANCE. The Lewis theory is valuable in explaining neutralization reactions of every kind, including those that:

(1) do not involve hydrogen,
(2) occur in nonaqueous solutions.

The Brønsted-Lowry Concept. This theory is less general than the Lewis theory since it requires that an acid be a hydrogen-containing substance. No such restriction is placed on a base.

ACID. An acid is a substance capable of donating protons. (It must, therefore, contain hydrogen.)

BASE. A base is a substance capable of accepting protons.

NEUTRALIZATION. Neutralization is the process of transferring a proton from an acid to a base. During such transfer a new acid is produced and the substance which lost the proton becomes a base, thus:

$$\underset{Acid_1}{HCl} + \underset{Base_1}{NH_3} \rightarrow \underset{Acid_2}{NH_4^+} + \underset{Base_2}{Cl^-}.$$

The acid, HCl, donates a proton to the base, NH_3, producing a new acid, NH_4^+, and a new base, Cl^-.

CONJUGATE PAIRS. The substance left behind after an acid donates a proton is the conjugate base of the acid. The base that acquires a proton becomes the conjugate acid of the original base. Thus in the equation above, the Cl^- is the conjugate base of HCl, NH_4^+ is the conjugate acid of NH_3, NH_3 is the conjugate base of NH_4^+, and HCl is the conjugate acid of Cl^-.

IMPORTANCE. The Brønsted-Lowry concept is valuable for considering water solutions and reactions in which hydrogen compounds act as acids. It emphasizes the fact that, in solutions, equilibria exist between acids and bases, and thus it is convenient to apply the law of mass action to neutralization (protolysis) reactions. The Brønsted-Lowry concept is especially valuable in dealing with reactions known as *hydrolysis*. Hydrolysis is a special case of protolysis. Cations such as Bi^{+3} and Sb^{+3} (see p. 130) or anions such as acetate and carbonate (see pp. 38–39 and 43) react with water to yield insoluble solids or slightly ionized substances. Considering hydrolysis as a simple protolysis (acid-base) reaction avoids the necessity of dealing with a separate concept.

Strong Acids and Bases

A strong acid or base is strong because it ionizes completely in dilute solution. In water solutions, acids form hydrated protons called hydronium ions, H_3O^+, and bases form hydroxide ions, OH^-. We shall deal with water solutions only, where neutralization always involves the formation of water, thus:

$$H_3O^+ + OH^- \rightleftharpoons 2H_2O. \tag{2–5}$$

Ion Product Constant for Water, K_w. In water there is some dissociation into H_3O^+ and OH^-, as is indicated by the double arrow in Equation (2–5). From the law of mass action (see p. 15) it can be shown that at any temperature the product of the molar concentrations of H_3O^+ and OH^- is a constant, thus:

$$[H_3O^+][OH^-] = K_w, \tag{2–6}$$

where brackets indicate molar concentrations, and K_w is the ion product constant for water.

TEMPERATURE AND K_w. K_w varies with the temperature. Thus at 25° C, K_w is 1.04×10^{-14}, but at 100° C it is 58.2×10^{-14}. This substantiates the generality that stabilities of substances decrease with increasing temperature.

THE pH SYSTEM. A convenient system of indicating the acidity of a solution is based upon K_w. This is the pH system, in which the pH is defined as the logarithm of the reciprocal of the hydronium ion concentration of the solution:

$$\text{pH} = \log \frac{1}{[H_3O^+]} = -\log [H_3O^+]. \tag{2–7}$$

In exactly the same way the hydroxide ion concentration can be expressed in terms of pOH:

$$\text{pOH} = \log \frac{1}{[OH^-]} = -\log [OH^-]. \tag{2–8}$$

In water, but only at 25° C:

$$\text{pH} = 14 - \text{pOH} \tag{2–9}$$

and:

$$\text{pOH} = 14 - \text{pH}. \tag{2–10}$$

Table 1 shows the relation of hydronium ion concentration to pH and pOH in water solutions at 25° C. A neutral solution has both a pH and a pOH of 7 at 25° C.

TABLE 1

Acid Solutions				Basic Solutions			
$[H_3O^+]$	pH	$[OH^-]$	pOH	$[H_3O^+]$	pH	$[OH^-]$	pOH
$1 = 10^0$	0	10^{-14}	14	10^{-14}	14	10^0	0
$0.1 = 10^{-1}$	1	10^{-13}	13	10^{-13}	13	10^{-1}	1
$0.01 = 10^{-2}$	2	10^{-12}	12	10^{-12}	12	10^{-2}	2
$0.001 = 10^{-3}$	3	10^{-11}	11	10^{-11}	11	10^{-3}	3
$0.0001 = 10^{-4}$	4	10^{-10}	10	10^{-10}	10	10^{-4}	4
$0.00001 = 10^{-5}$	5	10^{-9}	9	10^{-9}	9	10^{-5}	5
$0.000001 = 10^{-6}$	6	10^{-8}	8	10^{-8}	8	10^{-6}	6

At neutral: $[H_3O^+] = 10^{-7}$, pH = 7; $[OH^-] = 10^{-7}$, pOH = 7.

pH of Solutions of Strong Acids and Strong Bases. For strong acids and bases, if the solutions are dilute, the hydronium ion concentration can be considered equal to the normality of the acid, and the hydroxide ion concentration can be considered equal to the normality of the base. Thus a 0.01 N sodium hydroxide solution can be considered as having a hydroxide ion concentration of 0.01 molar.

Example 1. Calculate the pH of a 0.001 molar HCl solution.

Step 1. Find the $[H_3O^+]$.

If the HCl is completely dissociated, then the $[H_3O^+]$ is the same as the molar concentration of the acid, which is 0.001 M or 10^{-3} M.

Step 2. Calculate the pH. This is done by substituting in Equation (2–7):

$$\text{pH} = \log \frac{1}{10^{-3}} = \log 10^3 = 3.$$

Example 2. What is the pH of a 0.01 N solution of NaOH?

Step 1. Calculate the $[H_3O^+]$. This is done by substitution in Equation (2–6):

$$[H_3O^+][OH^-] = 10^{-14}.$$

$$[H_3O^+]0.01 = 10^{-14}.$$

$$[H_3O^+] = \frac{10^{-14}}{10^{-2}} = 10^{-12}.$$

Step 2. Calculate the pH, using Equation (2–7):

$$\text{pH} = \log \frac{1}{10^{-12}} = \log 10^{12} = 12.$$

An alternate method is:

Step 1. Calculate the $[OH^-]$. This is the same as the concentration of NaOH or 0.01 M or 10^{-2} M.

Step 2. Find the pOH, using Equation (2–8):

$$\text{pOH} = \log \frac{1}{10^{-2}} = \log 10^2 = 2.$$

Step 3. Calculate the pH, using Equation (2–9):

$$\text{pH} = 14 - 2 = 12.$$

Example 3. What is the pH of a 0.003 M HCl solution?

Step 1. Find the hydronium ion concentration.

The $[H_3O^+]$ is the same as the molar concentration of the HCl, which is 0.003 M = 3×10^{-3} M.

Step 2. Calculate the pH:

$$\text{pH} = \log \frac{1}{3 \times 10^{-3}} = \log 0.33 \times 10^3 = \log 3.3 \times 10^2 = 2.52.$$

To find the logarithm of 3.3×10^2, remember that the exponent of 10 is 2, and 2 is the logarithm of the number 10^2. The log of 3.3 is 0.52. To multiply, add the logarithms: $2 + 0.52 = 2.52$.

Calculation of pH during a Titration. The calculation of the pH of a solution during an acid-base titration gives results that can be plotted on a curve such as that in Fig. 1, with the pH plotted along one axis and the volume of base added to acid plotted along the other.

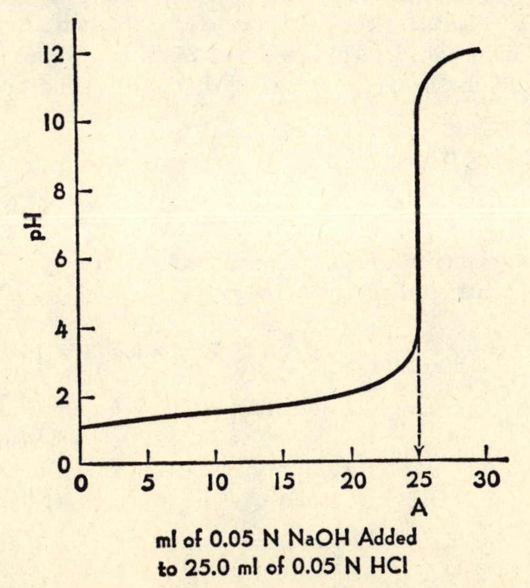

Fig. 1. Acid-Base Titration Curve.

The curve shown is a titration curve for a strong acid against a strong base. *A* is the equivalence point.

Example. During the titration of 25 ml of a 0.05 N HCl solution with 0.05 N NaOH solution, calculate (a) the pH after 24.00 ml of NaOH solution has been added, (b) the pH after 24.75 ml of NaOH has been added, and (c) the pH after 25.25 ml of NaOH has been added. Assume the total volume to be 50.00 ml after each addition.

(a) Step 1. Find the normality of the solution after the NaOH is added. Since the normalities are equal, 24 ml of NaOH will neutralize 24 ml of the original 25 ml of HCl solution. This leaves 1 ml of 0.05 N HCl not neutralized. Assuming the volume of the solution is 50 ml after the addition (it is really 49 ml), we calculate the normality from Equation (1–2) in Chapter 1:

$$V_1 \times N_1 = V_2 \times N_2.$$
$$1 \times 0.05 = 50 \times N_2.$$

Solving for N_2:
$$N_2 = \frac{0.05}{50} = 0.001 = 10^{-3}.$$

This is the normality of the solution with respect to HCl. It is also the molar concentration of H_3O^+ in the solution.

Step 2. Calculate the pH:

$$\text{pH} = \log \frac{1}{10^{-3}} = \log 10^3 = 3.$$

(b) Step 1. Calculate the normality of the solution after 24.75 ml of NaOH solution is added. There will be 0.25 ml of acid not neutralized in 50 ml of solution. Its normality and $[H_3O^+]$ are found by using Equation (1–2):

$$0.25 \times 0.05 = 50 \times N_2.$$
$$N_2 = \frac{0.05 \times 0.25}{50} = 0.00025 = 2.5 \times 10^{-4}.$$

This is the normality of the solution and also the $[H_3O^+]$.

Step 2. Calculate the pH:

$$\text{pH} = \log \frac{1}{2.5 \times 10^{-4}} = \log (0.4 \times 10^4) = \log (4 \times 10^3) = 3.60.$$

(c) Step 1. Calculate the normality of the solution after 25.25 ml of 0.05 N NaOH is added. This much NaOH solution will neutralize all the acid and leave an excess of 0.25 ml of NaOH solution, diluted to 50 ml. Its normality and $[OH^-]$ are found by using Equation (1–2):

$$0.25 \times 0.05 = 50 \times N_2.$$
$$N_2 = \frac{0.05 \times 0.25}{50} = 0.00025 = 2.5 \times 10^{-4}.$$

Step 2. Calculate the pOH:

$$\text{pOH} = \log\left(\frac{1}{2.5 \times 10^{-4}}\right) = \log(0.4 \times 10^{4}) = \log(4 \times 10^{3}) = 3.60.$$

Step 3. Calculate the pH, using Equation (2–9):

$$\text{pH} = 14 - 3.60 = 10.40.$$

Weak Acids and Bases

Whereas strong acids and bases are considered, when in solution, as existing in a completely ionized state, solutions of weak acids and bases contain not only positive and negative ions but molecules of the undissociated acids and bases. The relative amounts of ions and molecules are dependent upon the strength of the acid or base; the stronger the acid or base the larger the number of ions, relative to undissociated molecules; the weaker the acid or base the fewer the ions, relative to undissociated molecules. There is an equilibrium between ions and molecules.

An equilibrium constant is given by applying the law of mass action (see p. 15) to the reaction of ionization of a weak acid or base. Using acetic acid as an example, the reaction of ionization is:

$$HC_2H_3O_2 + H_2O \rightleftharpoons H_3O^+ + C_2H_3O_2^-.$$

The equilibrium constant for this reaction is:

$$K_E = \frac{[H_3O^+][C_2H_3O_2^-]}{[H_2O][HC_2H_3O_2]}. \tag{2–11}$$

But this equation contains $[H_2O]$, which in water solutions amounts to about $\frac{1000}{18} = 55.5$ moles per liter. This does not change appreciably from solution to solution, and its use merely complicates calculations. Since it never varies much, another equation can be set up, in terms of a new constant, which leaves out $[H_2O]$. This is the same as multiplying both sides of Equation (2–11) by 55.5. $K_E \times 55.5$ is called the *ionization constant* and is designated K_A for acids and K_B for bases. For acetic acid:

$$K_A = \frac{[H_3O^+][C_2H_3O_2^-]}{[HC_2H_3O_2]} = 1.8 \times 10^{-5}. \tag{2–12}$$

For the weak base NH_4OH, a corresponding ionization constant is:

$$K_B = \frac{[NH_4^+][OH^-]}{[NH_4OH]} = 1.8 \times 10^{-5}. \tag{2–13}$$

It is purely coincidental that the ionization constants for NH_4OH and for $HC_2H_3O_2$ are numerically equal. Values of K_A and K_B for several acids and bases are given in the table in Appendix V.

Calculation of pH from Ionization Constants. By the use of ionization constants the pH of solutions of acids and bases, of their salts, or of mixtures of acids or bases with their salts, can be calculated.

Example 1. Calculate the pH of a 0.10 N solution of acetic acid.

Step 1. Substitute all known values in the equation for the ionization constant for acetic acid, Equation (2–12). The chemical equation for the ionization of acetic acid shows that for every molecule of acid ionized, one acetate ion and one hydronium ion are produced. Therefore, in such a solution $[H_3O^+] = [C_2H_3O_2^-]$ and, as a result of this equality, in Equation (2–12), $[H_3O^+][C_2H_3O_2^-]$ is the same value as $[H_3O^+]^2$.

Substituting $[H_3O^+]^2$ for $[H_3O^+][C_2H_3O_2^-]$ in Equation (2–12), we get:

$$1.8 \times 10^{-5} = \frac{[H_3O^+]^2}{[HC_2H_3O_2]}.$$

Step 2. The acetic acid concentration is less than 0.10 N because of the fact that some acetic acid molecules are used up in forming ions. For $[HC_2H_3O_2]$ substitute $0.10 - [H_3O^+]$ to get a quadratic equation. It can be shown that the value obtained by solving this quadratic equation is very nearly the same as the one obtained by ignoring the decrease in $[HC_2H_3O_2]$ and solving Equation (2–12) with 0.10 substituted for the $[HC_2H_3O_2]$, thus:

$$1.8 \times 10^{-5} = \frac{[H_3O^+]^2}{0.10}.$$

The results are so similar because such a small portion of the acetic acid molecules, relative to the total number, is used in forming ions. Similar approximations are generally made throughout analytical chemical calculations.

Step 3. Solve for the molar concentration of hydronium ions:

$$[H_3O^+]^2 = 1.8 \times 10^{-5} \times 0.10.$$
$$[H_3O^+] = \sqrt{1.8 \times 10^{-6}} = 1.34 \times 10^{-3}.$$

Step 4. Calculate the pH, using Equation (2–7):

$$\text{pH} = \log \left[\frac{1}{1.34 \times 10^{-3}}\right] = \log [0.747 \times 10^3]$$
$$= \log [7.47 \times 10^2] = 2.87.$$

Example 2. Calculate the pH of a solution which is 0.10 N in $HC_2H_3O_2$ and 0.20 N in $NaC_2H_3O_2$.

Step 1. Substitute all known values in the equilibrium constant Equation (2–12):

$$1.8 \times 10^{-5} = \frac{[H_3O^+] \times 0.20}{0.10}.$$

Note. The 0.20 N $NaC_2H_3O_2$ yields a solution 0.20 M in $C_2H_3O_2^-$ by complete ionization. The $[C_2H_3O_2^-]$ is therefore taken as 0.20, although a minute amount of acetate ion is produced by ionization of acetic acid, $HC_2H_3O_2$. The 0.10 N $HC_2H_3O_2$ is thus assumed to remain entirely in the molecular form as 0.10 M acetic acid although a trace of it is ionized, yielding an increase in acetate ion concentration equal to the increase in hydronium ions. The values thus used for the calculation are therefore very close approximations, rather than exact quantities.

Step 2. Solve for the $[H_3O^+]$:

$$[H_3O^+] = \frac{1.8 \times 10^{-5} \times 0.10}{0.20} = 0.90 \times 10^{-5} = 9 \times 10^{-6}.$$

Step 3. Calculate the pH:

$$pH = \log \left[\frac{1}{9 \times 10^{-6}}\right] = \log [0.111 \times 10^{6}] = \log [1.11 \times 10^{5}] = 5.05.$$

The situation described in Example 2 may be created either by adding acetic acid and sodium acetate to water or by starting with acetic acid and neutralizing part of it with sodium hydroxide to produce some sodium acetate, but leaving some acid not neutralized. (For special properties of this type of solution see "Buffer Solutions," p. 36.)

Example 3. Calculate the pH of a solution which is 0.20 M in NH_4OH and 0.70 M in NH_4Cl.

Step 1. Substitute all known values in the ionization constant Equation (2–13) for ammonium hydroxide, making the same type of approximations as in Example 2:

$$1.8 \times 10^{-5} = \frac{0.70 \times [OH^-]}{0.20}.$$

Step 2. Solve for the OH^-:

$$[OH^-] = \frac{1.8 \times 10^{-5} \times 0.20}{0.70} = 0.514 \times 10^{-5} = 5.14 \times 10^{-6}.$$

Step 3. Solve for the pOH, using Equation (2–8):

$$pOH = \log \left(\frac{1}{5.14 \times 10^{-6}}\right) = \log (0.1944 \times 10^{6})$$
$$= \log (1.944 \times 10^{5}) = 5.29.$$

Step 4. Find the pH, using Equation (2–9):

$$pH = 14 - 5.29 = 8.71.$$

Buffer Solutions. Buffer solutions contain either a weak acid and some salt of the acid or a weak base and some salt of the base. It is possible to add considerable quantities of a strong acid or a strong base to a buffer solution with only a small change in the pH resulting.

Assuming we have a buffer solution containing acetic acid and sodium acetate, how does it maintain a nearly fixed pH? If a strong base such as NaOH solution is added to the buffer, resulting in additional hydroxide ions, the acetic acid molecules react thus:

$$HC_2H_3O_2 + OH^- \rightarrow H_2O + C_2H_3O_2^-.$$

If not all of the acetic acid is used up, the acetate ion concentration is increased exactly as much as the acetic acid concentration is decreased.

If a strong acid such as HCl is added, more hydronium ions enter the solution. These hydronium ions react with the acetate ions to form acetic acid molecules, thus:

$$C_2H_3O_2^- + H_3O^+ \rightarrow HC_2H_3O_2 + H_2O.$$

If not all of the acetate ions are used up, the acetic acid concentration is increased as much as the acetate ion concentration is decreased.

Such changes as are described above are changes in the ratio $\frac{[C_2H_3O_2^-]}{[HC_2H_3O_2]}$ which appears in the ionization constant equation, usually resulting in small changes in pH. This can be shown easily by a sample calculation.

Example. If 2 ml of 1.0 N solution of HCl is added to 100 ml of a solution that is 0.2 N in $NaC_2H_3O_2$ and 0.2 N in $HC_2H_3O_2$, what change in pH will occur? Assume 100 ml is the final volume.

Step 1. The pH before the addition of the acid must be found by substituting all known values in the ionization constant equation for acetic acid, Equation (2–12). Solving for the hydronium ion concentration and then for the pH, using Equation (2–7), gives the initial pH:

$$1.8 \times 10^{-5} = \frac{[H_3O^+]0.20}{0.20}.$$

$$[H_3O^+] = 1.8 \times 10^{-5}.$$

$$pH = \log \frac{1}{1.8 \times 10^{-5}} = \log 0.555 \times 10^5$$

$$= \log 5.55 \times 10^4 = 4.74, \text{ the initial pH.}$$

Step 2. Find the pH after the acid is added. This is done in several steps. First find the amount of the acetate ion in the 100 ml of buffer that was changed to acetic acid. Assume (although this is not absolutely true) that all the acid added as HCl converts an equivalent quantity of acetate ions to acetic acid molecules. The amount of the acetate converted is found by applying Equation (1–2):

$$2.0 \times 1.0 = V_2 \times 0.20.$$

Here 2.0 is the volume of HCl solution added, 1.0 is the normality of this solution, V_2 is the volume of 0.2 N acetate solution converted, and 0.20 is the normality of the original acetate solution. Solving for V_2:

$$V_2 = \frac{2.0 \times 1.0}{0.20} = 10.$$

10 ml of the 0.20 N acetate was converted to acetic acid by the HCl. This is one-tenth of the amount that was in the original 100 ml. Therefore the molarity of the acetate left is only nine-tenths its initial molarity.

$$0.20 \times \tfrac{9}{10} = 0.18 \text{ M}.$$

This is the molarity of the remaining acetate in the buffer solution.

The acetic acid concentration is increased proportionately to eleven-tenths its initial concentration.

$$0.20 \times \tfrac{11}{10} = 0.22 \text{ M}.$$

This is the molarity of the acetic acid in the buffer solution.

Substituting these values in the equation for the ionization constant for acetic acid, Equation (2–12), we get:

$$1.8 \times 10^{-5} = \frac{[H_3O^+]0.18}{0.22}.$$

$$[H_3O^+] = \frac{1.8 \times 10^{-5} \times 0.22}{0.18} = 2.2 \times 10^{-5}.$$

The pH is found from this by using Equation (2–7):

$$\text{pH} = \log \frac{1}{2.2 \times 10^{-5}} = \log 0.455 \times 10^5 = \log 4.55 \times 10^4 = 4.66.$$

Step 3. Find the change in pH by taking the difference in pH before and after the addition of acid:

$$4.74 - 4.66 = 0.08 \text{ pH units change.}$$

This is a very small change in pH with the addition of an appreciable quantity of strong acid. Addition of strong base changes the pH in an exactly similar manner, except that the pH becomes higher.

pH of a Solution of a Salt of a Weak Acid and a Strong Base. The solution of a salt of a weak acid and a strong base is the same as a solution in which chemically equivalent quantities of the acid and base have been added to water. In this type of solution the hydroxide ions tend to acquire hydronium ions to form slightly ionized water molecules. At the same time the negative ions of the salt tend to acquire hydronium ions which form the slightly ionized, weak acid. In a solution of a salt of acetic acid, for example, the following reactions would be in competition:

$$H_3O^+ + OH^- \rightleftharpoons 2H_2O.$$
$$H_3O^+ + C_2H_3O_2^- \rightleftharpoons HC_2H_3O_2 + H_2O.$$

In all solutions in which two such equilibria are both established in the same vessel, the $[H_3O^+]$ is always the same value in any equations involving either equilibrium. Thus in Equation (2–6):

$$[H_3O^+] = \frac{K_w}{[OH^-]},$$

but from Equation (2–12):

$$[H_3O^+] = \frac{K_A[HC_2H_3O_2]}{[C_2H_3O_2^-]}.$$

Since $[H_3O^+]$ can have but one value in the same beaker, the following equation would be true:

$$\frac{K_w}{[OH^-]} = \frac{K_A[HC_2H_3O_2]}{[C_2H_3O_2^-]}. \qquad (2\text{–}14)$$

But if we put sodium acetate into water, the only $HC_2H_3O_2$ molecules in the solution will come from the following hydrolysis reaction (see p. 28):

$$C_2H_3O_2^- + H_2O \rightarrow HC_2H_3O_2 + OH^-.$$

From this equation it can be seen that there will be as many acetic acid molecules as hydroxide ions in such a solution. Substituting $[OH^-]$ for $[HC_2H_3O_2]$ in Equation (2–14), we get:

$$\frac{K_w}{[OH^-]} = \frac{K_A[OH^-]}{[C_2H_3O_2^-]},$$

in which K_w is the ion product constant for water and K_A is the ionization constant for the weak acid, in this case acetic acid. Solving for $[OH^-]$:

$$[OH^-]^2 = \frac{K_w[C_2H_3O_2^-]}{K_A}.$$

$$OH^- = \sqrt{\frac{K_w[C_2H_3O_2^-]}{K_A}}. \qquad (2\text{–}15)$$

This is a general equation in which the concentration of any anion (negative ion) may be substituted for $[C_2H_3O_2^-]$. For practical purposes the amount of negative ion changed to the acid by reaction with water is negligible if K_A is 1×10^{-3} or less. Therefore the $[C_2H_3O_2^-]$ in solution is assumed to be the molar concentration calculated from the amount of salt added to the solution, provided solutions no more concentrated than about 0.2 M are involved.

In Equation (2–15) the value $\frac{K_w}{K_A}$ is often combined to form a new constant known as the hydrolysis constant, K_{Hyd}.

$$[OH^-] = \sqrt{K_{Hyd}\,[C_2H_3O_2^-]}.$$

Example. Find the pH of a solution of 0.20 N sodium acetate. Disregard the change of acetate ions to acetic acid molecules.

Step 1. Solve for the $[OH^-]$ after substituting in Equation (2–15):

$$\begin{aligned}[OH^-] &= \sqrt{\frac{10^{-14} \times 0.20}{1.8 \times 10^{-5}}} \\ &= \sqrt{\frac{2 \times 10^{-15}}{1.8 \times 10^{-5}}} \\ &= \sqrt{1.11 \times 10^{-10}} \\ &= 1.05 \times 10^{-5}.\end{aligned}$$

Step 2. Solve for the pOH of the solution, using Equation (2–8):

$$\text{pOH} = \log \frac{1}{1.05 \times 10^{-5}} = \log 0.95 \times 10^5 = \log 9.5 \times 10^4 = 4.98.$$

Step 3. Calculate the pH, using Equation (2–9):

$$\text{pH} = 14 - 4.98 = 9.02.$$

This is also the pH at the equivalence point during a titration of acetic acid with sodium hydroxide.

pH of a Solution of a Salt of a Weak Base and a Strong Acid. In the solution of the salt of a weak base and a strong acid, the positive ions of the salt compete with the hydronium ions of the water for hydroxide ions in the solution. The equations for the reactions, where NH_4Cl is taken as an example, are:

$$NH_4^+ + OH^- \rightleftharpoons NH_4OH$$

and:

$$H_3O^+ + OH^- \rightleftharpoons H_2O.$$

Slightly ionized substances are formed in both reactions. With exactly the same reasoning as was used for finding the $[OH^-]$ of a sodium acetate solution, the $[H_3O^+]$ for an ammonium chloride solution can be found thus:

$$[H_3O^+] = \sqrt{\frac{K_w[NH_4^+]}{K_B}}. \qquad (2\text{–}16)$$

This is a general equation in which the concentration of cations (positive ions) of a salt of any weak base may be substituted for the $[NH_4^+]$. K_w is the ion product constant for water, and K_B is the ionization constant for a weak base. The equation gives good values only for dilute solutions, 0.2 M or less.

As in Equation (2–15) the value $\frac{K_w}{K_A}$ may be combined to give the hydrolysis constant for a salt of a weak base, K_{Hyd}.

Example. Calculate the pH of a 0.20 N solution of hydrazine hydrochloride. K_B for hydrazine is 3×10^{-6}.

Step 1. Substitute all values known in Equation (2–16):

$$\begin{aligned}[H_3O^+] &= \sqrt{\frac{10^{-14} \times 0.20}{3 \times 10^{-6}}}\\ &= \sqrt{\frac{2 \times 10^{-15}}{3 \times 10^{-6}}}\\ &= \sqrt{0.667 \times 10^{-9}}\\ &= \sqrt{6.67 \times 10^{-10}}\\ &= 2.6 \times 10^{-5}.\end{aligned}$$

Step 2. Calculate the pH, using Equation (2–7):

$$\text{pH} = \log \frac{1}{2.6 \times 10^{-5}} = \log 0.385 \times 10^5 = \log 3.85 \times 10^4 = 4.59.$$

This is the pH at the equivalence point in titrating hydrazine with hydrochloric acid.

pH of the Salt of a Weak Base and a Weak Acid. Using reasoning similar to that used for deriving Equation (2–15), the following equation can be obtained. It, like Equations (2–15) and (2–16), is an approximation that is very close to the exact value:

$$[H_3O^+] = \sqrt{\frac{K_A K_w}{K_B}}. \qquad (2\text{–}17)$$

K_A is the ionization constant for a weak acid, K_B is the ionization constant for a weak base, and K_w is the ion product constant for water.

Example 1. Calculate the pH of a 0.10 N solution of $NH_4C_2H_3O_2$.

Step 1. Substitute all known values in Equation (2–17); K_A for acetic acid is 1.8×10^{-5}, K_B for NH_4OH is 1.8×10^{-5}, and K_w is 1×10^{-14}:

$$[H_3O^+] = \sqrt{\frac{1.8 \times 10^{-5} \times 1 \times 10^{-14}}{1.8 \times 10^{-5}}}.$$

Step 2. Solve for $[H_3O^+]$:

$$[H_3O^+] = \sqrt{1 \times 10^{-14}} = 1 \times 10^{-7}.$$

Step 3. Calculate the pH of the solution, using Equation (2–7):

$$\text{pH} = \log \frac{1}{1 \times 10^{-7}} = \log 10^7 = 7.$$

This is the pH at the point where equivalent quantities of the acid and base have been mixed.

Example 2. What is the pH of the solution of the salt formed when hydrazine is neutralized with acetic acid? K_A for acetic acid $= 1 \times 10^{-5}$, K_B for hydrazine $= 3 \times 10^{-6}$.

Step 1. Substitute known values in Equation (2–17) and solve for the $[H_3O^+]$:

$$\begin{aligned}[H_3O^+] &= \sqrt{\frac{1.8 \times 10^{-5} \times 1 \times 10^{-14}}{3 \times 10^{-6}}} \\ &= \sqrt{6 \times 10^{-14}} \\ &= 2.45 \times 10^{-7}.\end{aligned}$$

Step 2. Calculate the pH, using Equation (2–7):

$$\begin{aligned}\text{pH} &= \log \frac{1}{2.45 \times 10^{-7}} = \log 0.408 \times 10^7 \\ &= \log 4.08 \times 10^6 = 6.61.\end{aligned}$$

This is the pH of a solution of a salt of hydrazine with acetic acid.

Polyequivalent Acids and Their Salts

Such acids as H_2S, H_2SO_3, H_2CO_3, and H_3PO_4 are able to ionize, giving up 2 or more hydronium ions per mole of acid. These acids and others like them are polyequivalent acids, and they play an important role in qualitative analysis. If we use H_2S as a typical polyequivalent acid, it can be seen that ionization takes place in two steps:

Step 1. $$H_2S + H_2O \rightleftharpoons H_3O^+ + HS^-. \tag{2–18}$$

The ionization constant for this first step is designated as K_1 and is:

$$K_1 = \frac{[H_3O^+][HS^-]}{[H_2S]} = 5.9 \times 10^{-8}. \tag{2–19}$$

Step 2. $$HS^- + H_2O \rightleftharpoons H_3O^+ + S^{-2}. \quad (2\text{–}20)$$

The ionization constant for this step is designated as K_2 and is:

$$K_2 = \frac{[H_3O^+][S^{-2}]}{[HS^-]} = 1 \times 10^{-15}. \quad (2\text{–}21)$$

The values for K_1 and K_2, 5.9×10^{-8} and 1×10^{-15} respectively, are taken from Appendix V.

Calculation of the pH of a Solution of H_2S. This is valuable as an example which can be applied only to those polyequivalent acids with the first ionization constant 1000 or more times as large as the second. The first ionization constant is employed as if the acid were a monoequivalent acid such as acetic acid. The second ionization constant is so much smaller than the first that its effect can be disregarded.

Example. Calculate the pH of a 0.05 molar H_2S solution.

$$K_1 = 5.9 \times 10^{-8} \quad \text{and} \quad [H_3O^+] \cong [HS^-]$$

where $\cong$ means "approximately equal to."

Step 1. Substitute known values in Equation (2–19) and solve for the $[H_3O^+]$:

$$K_1 = \frac{[H_3O^+][HS^-]}{[H_2S]} = 5.9 \times 10^{-8}.$$

Since $[H_3O^+]$ is approximately equal to $[HS^-]$, we can rewrite the equation thus:

$$\frac{[H_3O^+][H_3O^+]}{[H_2S]} = 5.9 \times 10^{-8}$$

and solve this for the $[H_3O^+]$:

$$\begin{aligned} [H_3O^+]^2 &= 5.9 \times 10^{-8} \times 0.05. \\ [H_3O^+] &= \sqrt{0.295 \times 10^{-8}} \\ &= \sqrt{29.5 \times 10^{-10}} \\ &= 5.43 \times 10^{-5}. \end{aligned}$$

Step 2. Substitute this value of $[H_3O^+]$ in Equation (2–7) and solve for the pH:

$$\begin{aligned} \text{pH} &= \log \frac{1}{5.43 \times 10^{-5}} = \log 0.184 \times 10^5 \\ &= \log 1.84 \times 10^4 = 4.26. \end{aligned}$$

Calculation of the pH of Solutions of Salts of Polyequivalent Acids. The salt formed when 1 equivalent weight of a strong base and 1 formula weight of a diequivalent acid react is an acid salt, a hydrogen salt, or sometimes a "bisalt." NaHS is sodium hydrogen sulfide; $NaHCO_3$ is sodium hydrogen carbonate or sodium bicarbonate. A rather involved derivation gives an approximate hydronium ion concentration for a solution of such a salt:

$$[H_3O^+] = \sqrt{K_1K_2}, \qquad (2\text{–}22)$$

where K_1 and K_2 are the two ionization constants for the acid.

Example. Calculate the pH of a sodium hydrogen sulfide solution. All but very dilute and very concentrated solutions will have approximately the same pH.

Step 1. Substitute known values in Equation (2–22) and solve for the hydronium ion concentration:

$$\begin{aligned}[H_3O^+] &= \sqrt{5.9 \times 10^{-8} \times 1 \times 10^{-15}} \\ &= \sqrt{5.9 \times 10^{-23}} \\ &= \sqrt{59 \times 10^{-24}} \\ &= 7.68 \times 10^{-12}.\end{aligned}$$

Step 2. Calculate the pH using Equation (2–7):

$$\begin{aligned}pH &= \log \frac{1}{7.68 \times 10^{-12}} = \log 0.13 \times 10^{12} \\ &= \log 1.3 \times 10^{11} = 11.11.\end{aligned}$$

Calculation of the pH of a Solution of a Normal Salt of a Diequivalent Acid. A salt formed by adding 2 equivalent weights of a strong base to 1 formula weight of a diequivalent acid is called the *normal salt.* The pH of its solution is found by reasoning that only one ionization constant, the second, need be considered in calculating the pH of the solution. This assumption is reasonably correct when K_1 and K_2 differ from each other greatly, as with H_2S and H_2CO_3. It is not a safe assumption where K_1 is less than 100 times as large as K_2.

Example. Calculate the pH of a 0.1 M solution of Na_2CO_3.

Step 1. Apply Equation (2–15) by substituting known values and inserting only the value of K_2 for H_2CO_3 as K_A and solving for the hydroxide ion concentration. The equilibrium is one known as "hydrolysis" (see p. 28) and the equation is $CO_3^{-2} + H_2O \rightleftharpoons HCO_3^- + OH^-$:

$$\begin{aligned}[OH^-] &= \sqrt{\frac{1 \times 10^{-14} \times 0.1}{4.2 \times 10^{-11}}} \\ &= \sqrt{\frac{1 \times 10^{-15}}{4.2 \times 10^{-11}}} \\ &= \sqrt{0.238 \times 10^{-4}} \\ &= \sqrt{23.8 \times 10^{-6}} \\ &= 4.88 \times 10^{-3}.\end{aligned}$$

Step 2. Solve for the pOH, using Equation (2–8):

$$pOH = \log \frac{1}{4.88 \times 10^{-3}} = \log 0.205 \times 10^{3}$$
$$= \log 2.05 \times 10^{2} = 2.31.$$

Step 3. Solve for the pH by applying Equation (2–9):

$$pH = 14 - 2.31 = 11.69.$$

It should be understood that such a calculation gives only an approximation because approximations are made in deriving Equation (2–15) and K_1 is ignored.

Calculation of the Sulfide Ion Concentration in Acid Solutions. Sulfide ions are used to precipitate various metal ions as sulfides from solutions in the qualitative analysis procedure. By adjustment of the concentration of hydronium ions the concentration of sulfide ions can be controlled closely enough to allow sulfides of some metals to precipitate while others do not precipitate. Thus a separation of one group of metal ions from another is effected.

Equations (2–18) and (2–20),

$$H_2S + H_2O \rightleftharpoons H_3O^+ + HS^-$$

and

$$HS^- + H_2O \rightleftharpoons H_3O^+ + S^{-2},$$

show that the second step in the ionization of H_2S is dependent on the first step. That is, no S^{-2} can be formed from H_2S without HS^- being formed first by the reaction shown in Equation (2–18). Thus if we add the two equations, getting

$$H_2S + 2H_2O \rightleftharpoons 2H_3O^+ + S^{-2},$$

the equilibrium constant, K_A, for the combined reactions must be a product of K_1 and K_2:

$$K_A = K_1 \times K_2.$$
$$K_A = \frac{[H_3O^+]^2[S^{-2}]}{[H_2S]} = 5.9 \times 10^{-8} \times 1 \times 10^{-15}$$
$$= 5.9 \times 10^{-23}. \qquad (2\text{–}23)$$

The solubility of H_2S in water and acid solutions is about 0.1 molar. Equation (2–23) can then be applied to find the sulfide ion concentration in a solution of any hydronium ion concentration, if the solution is simultaneously saturated with H_2S. The hydronium ion concentration is adjusted to about 0.2–0.5 M by the addition of HCl solution to effect precipitation of the copper group sulfides without precipitating the nickel group sulfides.

Example. Calculate the sulfide ion concentration in a solution saturated with H_2S and 0.3 M in HCl. The HCl is completely ionized, giving a 0.3 M solution of H_3O^+.

Substitute known values in Equation (2–23) and solve for the sulfide ion concentration:

$$\frac{0.3^2 \times [S^{-2}]}{0.1} = 5.9 \times 10^{-23}.$$

$$0.09 \times [S^{-2}] = 5.9 \times 10^{-23} \times 0.1.$$

$$[S^{-2}] = \frac{5.9 \times 10^{-24}}{0.09}$$

$$= 65.56 \times 10^{-24}$$

$$= 6.56 \times 10^{-23}.$$

Calculation of the Solubility of Metal Sulfides in H_2S Solution. Once the sulfide ion concentration is known, one can apply the solubility product constant (p. 16) to calculate the concentration of a metal ion that will saturate a solution with the sulfide of the metal. The sulfide of the metal must be very slightly soluble in water.

Example 1. Calculate the molar concentration of Cu^{+2} that will saturate a solution with CuS if the solution is saturated with H_2S and the $[H_3O^+]$ is 0.3 M. The solubility product constant for CuS is 9×10^{-36}.

Substitute the value for the sulfide ion concentration in the solubility product equation and solve for the copper ion concentration:

$$[Cu^{+2}] \times 6.56 \times 10^{-23} = 9 \times 10^{-36}.$$

$$[Cu^{+2}] = \frac{9 \times 10^{-36}}{6.56 \times 10^{-23}}$$

$$= 1.37 \times 10^{-13}.$$

Thus the copper ion concentration can be reduced to 1.37×10^{-13} moles per liter by H_2S in a 0.3 M acid solution. This is an extremely small concentration.

Example 2. Calculate the concentration of zinc ions in moles per liter that will saturate a solution with ZnS in a 0.3 M solution of HCl ($[H_3O^+] = 0.3$) if the solution is saturated with H_2S. Again the sulfide ion concentration is 6.56×10^{-23} molar. The solubility product constant for ZnS is 1.0×10^{-21}.

Substitute the sulfide ion concentration in the solubility product equation for ZnS and solve for the zinc ion concentration:

$$[Zn^{+2}] \times 6.56 \times 10^{-23} = 1.0 \times 10^{-21}.$$

$$[Zn^{+2}] = \frac{1.0 \times 10^{-21}}{6.56 \times 10^{-23}} = 0.152 \times 10^2 = 15.2 \text{ molar}.$$

The concentration of a zinc compound would need to be 15.2 molar in order to saturate such a solution with ZnS. No ordinary salt of zinc is that soluble.

The two examples show why copper ions can be separated from zinc ions by H_2S precipitation of the copper sulfide from a 0.3 M HCl solution.

Acid-Base Indicators

Acid-base, or neutralization, indicators are colored materials, often dyes, which are one color in one pH range and another color, or colorless, in a different pH range. All acid-base indicators are either weak acids or weak bases with a characteristic ionization constant value. Therefore they show a color change at a pH which is characteristic for each. It is clear that changes in color are due to changes from ions to molecules, or the reverse.

General Indicator Reactions. For an acid type of indicator the following equation represents the chemical change that causes the color change:

$$\underset{\text{Color } A}{\text{HInd}} + H_2O \rightleftharpoons H_3O^+ + \underset{\text{Color } B}{\text{Ind}^-}. \qquad (2\text{–}24)$$

"HInd" represents the molecular form as an acid with color *A*. "Ind^-" represents the ion derived from the acid, the ion having color *B*.

For a base type of indicator:

$$\underset{\text{Color } A}{\text{IndOH}} \rightleftharpoons \underset{\text{Color } B}{\text{Ind}^+} + OH^-. \qquad (2\text{–}25)$$

"IndOH" represents the molecular form of the base indicator. "Ind^+" represents the ion derived from the base.

For either acid or base indicators, colors *A* and *B* may be two different colors, or one may be colorless.

Ionization Constants for Indicators. For each reversible reaction such as is indicated above, an ionization constant, K_{Ind}, can be found. For an acid indicator, as an example:

$$K_{\text{Ind}} = \frac{[H_3O^+][\text{Ind}^-]}{[\text{HInd}]}. \qquad (2\text{–}26)$$

The numerical value of K_{Ind} for either an acid or a base is seldom applied as such, even though its magnitude determines the pH at which a color change will occur.

pH of Indicator Color Change. Rearranging Equation (2–26) thus:

$$[H_3O^+] = K_{\text{Ind}} \frac{[\text{HInd}]}{[\text{Ind}^-]},$$

it becomes clear that where [HInd] = [Ind$^-$], the hydronium ion concentration becomes equal to the numerical value of K_{Ind}. At this point half the indicator is in the ionic state and the other half is in the molecular state. The indicator is changing color at the fastest rate during a titration, per unit of volume of reagent being added, and its color will be neither *A* nor *B*, but halfway between. In addition:

$$pH = pK_{Ind}.$$

This is true because at this point $[H_3O^+] = K_{Ind}$ and:

$$\log \frac{1}{[H_3O^+]} = \log \frac{1}{K_{Ind}}.$$

Range of Color Change of Indicators. Along with properties of indicators one generally finds listed a range of pH through which the indicator changes color. The pH of the middle of the range listed is approximately equal to the p*K* of the indicator. A list of indicators and the range of pH of the color change for each is given in Appendix IV.

From the rearrangement of Equation (2–26) to:

$$[H_3O^+] = K_{Ind} \frac{[HInd]}{[Ind^-]}$$

it is obvious that increasing the value of the ratio $\frac{[HInd]}{[Ind^-]}$ from 1 to 10 decreases the pH one unit. Changing this same ratio from 1 to 0.1 increases the pH one unit. Experiment has shown that when the intensity of one color of the indicator reaches about 10% of that of the other color, a change in color is detectable. Thus most indicators have a range of visible change of about 2 pH units, from one above to one below the point of equal concentrations of both colors of the indicator.

Effect of Temperature. Temperature affects the value of K_w and also the K_{Ind} for indicators. The indicator should be calibrated if it is to be used at other than near-room temperature.

Effect of Solvent. Solvents other than water may cause alteration in the value of K_{Ind}. If nonaqueous solvents are involved, the indicator should be calibrated for the conditions of the analysis.

Review Questions and Problems

1. Write the equation for the law of mass action applied to the general reaction $A + 2B \rightarrow C + 3D$.
2. The solubility of silver iodide, AgI, is 2.35×10^{-6} g per liter in water. Calculate its molar solubility and its solubility product constant.
3. The solubility of lead chloride, $PbCl_2$, is 8.15 g per liter in water. Calculate its molar solubility and solubility product constant.
4. The solubility product constant for lead iodide, PbI_2, is 1.4×10^{-8}. Calculate its solubility in moles per liter and in grams per liter.
5. The solubility product for magnesium carbonate, $MgCO_3$, is 4×10^{-5}. What is its solubility in grams per 100 ml?
6. Write out the names for the following: $Co(NH_3)_4Cl^+$, $Co(NH_3)_4Cl_2$, $Pt(NH_3)_2Cl_2$, $Fe(C_2O_4)_3^{-3}$ [$C_2O_4^{-2}$ is the oxalate ion], AlF_6^{-3}, $Fe(CN)_6^{-3}$, $Cu(CN)_3^{-2}$, $Fe(CN)_6^{-4}$.
7. What two types of bonding exist in complex ions?
8. Calculate the silver ion concentration that would exist in solution if sufficient AgCl were dissolved in a 3 M NH_3 solution to produce a 0.001 M concentration of $Ag(NH_3)_2^+$.
9. What concentration of chloride ion would be required in the solution described in problem 8 to saturate the solution with AgCl? K_{sp} for AgCl is 1.2×10^{-10}.
10. What concentration of iodide would be required in the solution described in problem 8 to saturate the solution with AgI? K_{sp} for AgI is 1×10^{-16}. Problems 8–10 indicate why AgCl will dissolve in ammonia solution while AgI will not.
11. Define the terms *acid*, *base*, and *neutralization* according to both the Lewis electron-pair concept and the Brønsted-Lowry concept.
12. How does a strong acid or base differ from a weak acid or base?
13. Calculate the pH of:
    a. A 0.002 N solution of HCl.
    b. A 0.05 N solution of NaOH.
    c. A solution of 50 ml of 0.2 N NaOH to which 25 ml of 0.2 N HCl has been added. Total volume is 75 ml.
14. Find the pH of a 0.01 N solution of lactic acid. K_A for lactic acid is 1.4×10^{-4}.
15. K_A for hydrofluoric acid, HF, is 1.7×10^{-5}. Calculate the pH of a solution that is 0.2 N with respect to HF and 0.03 N with respect to NaF.
16. If to 100 ml of the solution in problem 15, 1 ml of 1.0 N NaOH were added, what would the pH be?
17. The K_B of aniline is 2.70×10^{-10}. What is the pH of a 0.1 N solution of the salt aniline hydrochloride?

18. What is the pH of a 0.01 N solution of potassium hydrogen phthalate? $K_1 = 1.29 \times 10^{-3}$, $K_2 = 3.8 \times 10^{-6}$. The measured value is about pH 4.12. Compare this with the calculated value.
19. Calculate the sulfide ion concentration of a solution saturated with H_2S, $[H_2S] = 0.1$, which is 0.2 M in HCl. K_A for H_2S is 5.9×10^{-23}.
20. In the solution in problem 19, calculate the nickel ion concentration needed to saturate the solution with NiS. K_{sp} for NiS is 2×10^{-21}.
21. Calculate the molar concentration of Bi^{+3} in a solution saturated with Bi_2S_3 and H_2S at a pH of 2. $[H_2S] = 0.1$. K_{sp} for $Bi_2S_3 = 7 \times 10^{-97}$.
22. Calculate the Bi^{+3} concentration in the same solution described in problem 21 except that the $[H_3O^+]$ is 0.3.
23. What conditions must exist if the $[H_3O^+] = K_{Ind}$?
24. Explain why the total range of visible color change of an indicator is generally about 2 pH units.

3

Analysis of Group I, the Silver Group

In the qualitative analysis of the metal ions certain operations are performed in a specific order. A solution of the sample is prepared. The groups are precipitated out one after the other. The group precipitates are filtered or centrifuged, washed free of the mother liquor, and analyzed for each possible metal that might be present. Chart 1 (p. 3) indicates that the first group of metal ions precipitated is called the *silver group*. The steps in the analysis of Group 1 are indicated in Chart 2. The operations are described below.

The procedures described here are on a rather small scale, called *semimicro* procedures, wherein precipitates are centrifuged out of suspension rather than being filtered away from a liquid. If 10 or 20 times as large quantities are used, the procedures are called *macro* procedures and the precipitates are filtered rather than centrifuged. Large volumes of liquids and amounts of precipitates are not centrifuged because they would require a very large, expensive centrifuge.

Step 1. Precipitation of Group I. If a sample is not in solution ready for analysis, it must be prepared for analysis by the steps given on pp. 120 or 127.

Using a medicine dropper, place 5 drops of the solution of the sample in a small test tube. Add 15 drops of distilled water, then 2 drops of 6 M HCl solution. Stir and centrifuge. When centrifuging, remember to balance the test tube with another containing an equal volume of water and placed on the opposite side of the centrifuge (see Fig. 2, p. 52). The precipitate should settle rapidly. Remove the test tubes from the centrifuge carefully so as not to stir up the precipitate.

CHART 2

Analysis of Group I

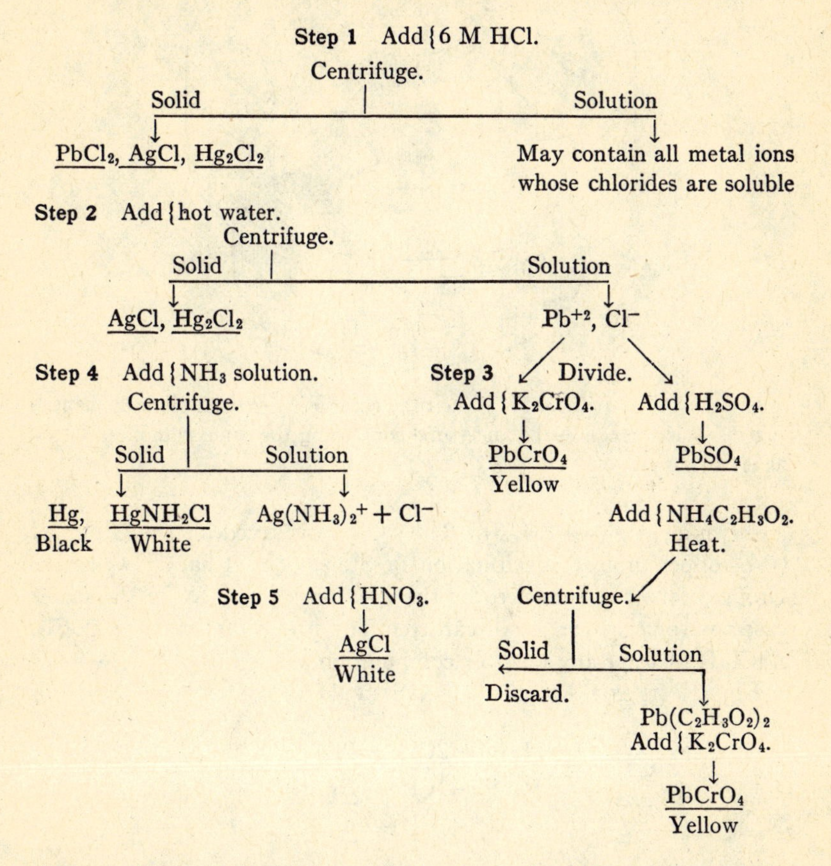

Allow 1 drop of 6 M HCl to trickle down the side of the test tube containing the precipitate. If the solution above the precipitate remains clear, or nearly so, all of the ions of Group I are precipitated. If more precipitate forms, insufficient precipitating agent (HCl solution) has been added. In that case, stir, centrifuge, and add another drop of 6 M HCl, and repeat until no precipitate forms on the addition of more HCl. Cool under running water and decant the clear solution from above the precipitate. Decanting on a semimicro scale is the process of removing the clear solution with a medicine dropper with

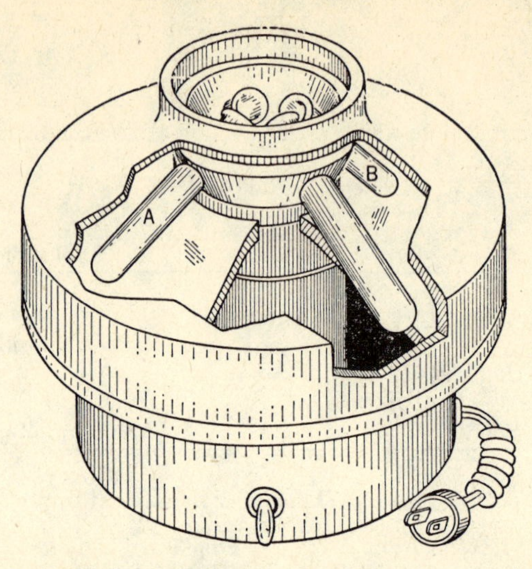

Fig. 2. A Centrifuge. A is the tube containing the sample to be centrifuged. B is the counterbalance tube containing the same volume of liquid as is in tube A.

a drawn-out tip (see Fig. 3). The solution is saved for the Group II (the copper group) precipitation in Step 6 (see Chapter 4) unless Group I ions are known to be the only ones present. In the latter case the solution can be discarded. The precipitate may be $PbCl_2$, $AgCl$, Hg_2Cl_2, or any possible combination of the three.

EQUATIONS FOR REACTIONS IN STEP 1:

$$Pb^{+2} + 2Cl^- \rightarrow \underline{PbCl_2}.$$

$$Hg_2^{+2} + 2Cl^- \rightarrow \underline{Hg_2Cl_2}.$$

$$Ag^+ + Cl^- \rightarrow \underline{AgCl}.$$

If a large amount of HCl solution as a precipitating agent is added in excess of that needed, the reaction

$$AgCl + Cl^- \rightarrow AgCl_2^-$$

occurs. $AgCl_2^-$ is soluble, and not all of the silver will be found in the precipitate.

It should also be noted that $PbCl_2$ is somewhat soluble in cold water and will precipitate in Group I only if present in rather large amounts.

The solution from Step 1 will contain lead ions which will appear in Group II if lead is present.

Step 2. Separation of $PbCl_2$ from $AgCl$ and Hg_2Cl_2. Advantage is taken of the fact that $PbCl_2$ is much more soluble than $AgCl$ and Hg_2Cl_2 in hot water.

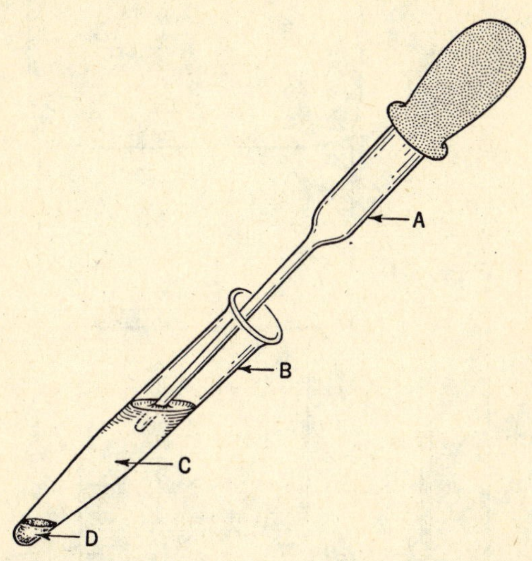

Fig. 3. Decanting on a Semimicro Scale. Decant the solution by drawing it up slowly into the drawn-out medicine dropper, called a "pipette." The tip of the pipette is held near the surface of the liquid and is lowered as the level of liquid is lowered. The last drop of liquid is taken up as slowly as possible so as not to disturb the precipitate in the bottom of the tube. A is the medicine dropper, B is a centrifuge tube, C is the solution to be decanted, and D is the precipitate.

To the precipitate in the test tube add 1 ml (20 drops) of distilled water. Heat the test tube for 2 minutes in a hot-water bath such as is shown in Fig. 4, with the beaker, D, half filled with water brought to boiling. Stir several times during the heating. Prepare a counterbalancing test tube, quickly remove the hot test tube from the water bath, and centrifuge no more than 30 seconds. Quickly remove the test tube from the centrifuge and decant (with a pipette) the clear solution. Save the solution for testing for lead in Step 3. Save the precipitate for Step 4.

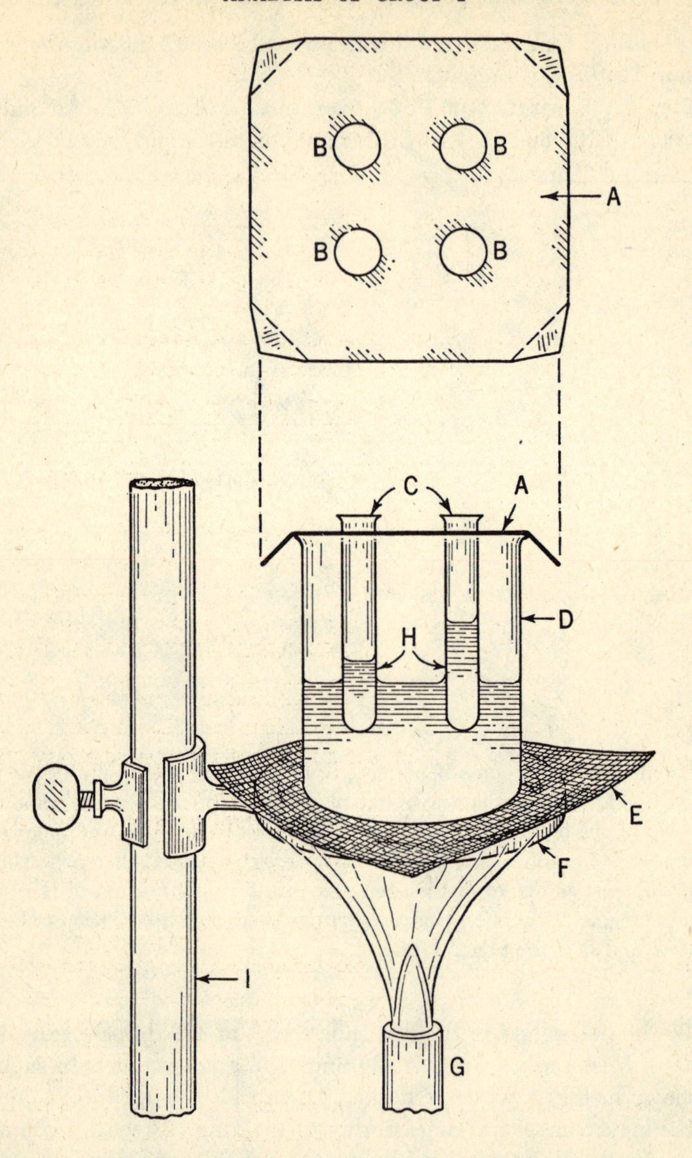

Fig. 4. This drawing shows a hot-water bath with the cover of the beaker, an aluminum or lead sheet, A, projected upward to give a view from above to show the holes, B, cut in the cover for tubes. C is one or more test tubes or centrifuge tubes. D is the beaker, E the screen, F a ring, G a burner, H the liquid being heated, and I a stand.

EQUATION FOR THE REACTION IN STEP 2:

$$PbCl_2 \rightarrow Pb^{+2} + 2Cl^-.$$

Step 3. Testing for Lead. Place 5 drops of the solution from Step 2 in another test tube. Add 10 drops of distilled water and 2 drops of 0.5 M K_2CrO_4 solution. If a yellow precipitate forms, that indicates that lead may be present in the sample. But since chromates of most metals are insoluble and are nearly all yellow, this test is not conclusive. To make certain of the presence — or absence — of lead, add 6 drops of 9 M H_2SO_4 (concentrated sulfuric acid diluted one-to-one with water by pouring acid into water) to the solution remaining from Step 2. Formation of a white, grainy precipitate (lead sulfate, $PbSO_4$) is excellent evidence that lead is present. This precipitate can be proved to be lead sulfate by centrifuging, decanting, adding 3 M ammonium acetate solution to the precipitate, and heating in a water bath. If the precipitate does not completely dissolve, decant the clear solution. Add 3 drops of 0.5 M K_2CrO_4 solution to the clear decantate. If a yellow precipitate ($PbCrO_4$) forms, the presence of lead in the sample is confirmed.

EQUATIONS FOR REACTIONS IN STEP 3:

$$Pb^{+2} + CrO_4^{-2} \rightarrow \underline{PbCrO_4}.$$

$$Pb^{+2} + SO_4^{-2} \rightarrow \underline{PbSO_4}.$$

$$PbSO_4 + 2C_2H_3O_2^- \rightarrow Pb(C_2H_3O_2)_2 + SO_4^{-2}.$$

$$Pb(C_2H_3O_2)_2 + CrO_4^{-2} \rightarrow \underline{PbCrO_4} + 2C_2H_3O_2^-.$$

The reaction between acetate ions in the ammonium acetate solution and solid lead sulfate occurs because the product, lead acetate, is a very slightly ionized substance. It is so slightly ionized that in a 3 M solution of acetate ions, fewer lead ions exist in solution than are present in a saturated solution of lead sulfate. However, there are sufficient lead ions in solution so that when chromate ions are added the solubility product constant for lead chromate is exceeded and lead chromate precipitates. Obviously lead chromate is much less soluble than lead sulfate or this test would not succeed.

Step 4. Separation of AgCl from Hg_2Cl_2 and Testing for the Presence of Mercurous Mercury. AgCl is separated from Hg_2Cl_2 by dissolving it in ammonia solution, forming the soluble $Ag(NH_3)_2^+$, a complex ion (see p. 20). At the same time the mercurous chloride reacts with NH_3 to form very small particles of Hg (which appear black) and $Hg(NH_2)Cl$ (white). The black predominates, so that the precipitate appears very dark gray, or black. This reaction of mercurous chloride with ammonia is known as an "auto-oxidation reaction" because Hg_2^{+2} reduces and oxidizes itself, resulting in metallic Hg and the mercuric amido chloride. See p. 12.

Wash the precipitate from Step 2 twice by stirring into it 1 ml of hot distilled water; then centrifuge both times, and discard the solution. This procedure washes lead chloride from the precipitate so that the addition of ammonia solution will not produce $Pb(OH)_2$ or $Pb(OH)Cl$ (both white), which might hide the black color of the mercury.

To the washed precipitate add 5 drops of distilled water and an equal volume of concentrated NH_3 solution (ammonium hydroxide). Stir, centrifuge, and decant the solution into a small beaker. Save the solution for the test for silver in Step 5. A dark gray or black precipitate remaining in the test tube from the ammonia treatment is proof of the presence of mercurous ions in the sample. A white precipitate or none at all indicates that no mercurous ions were present in the sample.

EQUATIONS FOR REACTIONS IN STEP 4:

$$AgCl + 2NH_3 \rightarrow Ag(NH_3)_2^+ + Cl^-.$$
$$Hg_2Cl_2 + 2NH_3 \rightarrow \underline{Hg} + \underline{HgNH_2Cl} + NH_4^+ + Cl^-.$$

Step 5. Testing for Silver. The solution from Step 4 may contain $Ag(NH_3)_2^+$ and Cl^-, along with a relatively large amount of ammonia. It is placed in a beaker rather than in a test tube, to prevent the large amount of heat developed as NH_3 is neutralized by HNO_3 from boiling the solution out of the container.

To the solution in the beaker add 3 M nitric acid slowly down the side of the beaker, stirring constantly to avoid spattering from boiling. After about 1 ml of acid is added and mixed, test the solution with litmus by removing the stirring rod and touching it to a piece of litmus paper. If the solution is not acid, add more acid and stir until it is completely acidified. If a white precipitate forms, it is AgCl, which proves that silver was present in the sample. If no precipitate forms, the absence of silver is indicated. The precipitate will first form at the point where the acid contacts the solution, and then it may redissolve in the remaining ammonia solution. Therefore it is necessary to continue to add acid, stir, and test with litmus until *all* of the solution is acid.

EQUATIONS FOR REACTIONS IN STEP 5:

$$Ag(NH_3)_2^+ + 2H_3O^+ \rightarrow Ag^+ + 2NH_4^+ + 2H_2O.$$
$$Ag^+ + Cl^- \rightarrow \underline{AgCl}.$$

Review Questions and Problems

1. Write out Chart 2 from memory.
2. From Chart 2 write the equations for every reaction occurring in the Group I analysis.
3. Name two compounds other than HCl which might be used to precipitate Group I ions.
4. After adding 0.1 ml of the precipitating agent (6 M HCl solution) in Step 1, the volume is 1.1 ml. Calculate the chloride and the H_3O^+ concentrations if no precipitate forms.
5. From the answer in problem 4, what concentration of Ag^+ and Pb^{+2} would just saturate the solution with AgCl and $PbCl_2$ respectively? K_{sp} for AgCl is 1.2×10^{-10}. K_{sp} for $PbCl_2$ is 1×10^{-4}. Assume that no complex ions such as $AgCl_2^-$ or $PbCl_4^{-2}$ are formed.
6. Write the ionic formula of a substance which will:
   a. Form a precipitate with a solution of either KCl or $ZnCl_2$.
   b. Form a precipitate with CrO_4^{-2} and also with Cl^-.
   c. Form a precipitate with HCl solution but not with NaCl solution.
   d. Form a precipitate with HCl solution and also with HNO_3 solution
   e. Dissolve $PbCl_2$ but not AgCl.
   f. Dissolve AgCl but not Hg_2Cl_2.
   g. Dissolve $PbSO_4$ but not $PbCrO_4$.
   h. Dissolve $Pb(NO_3)_2$ but not $PbSO_4$.
7. If too much concentrated HCl were added in precipitating Group I ions, which ion might not precipitate?
8. If H_2SO_4 solution rather than HCl solution were used to precipitate Group I ions, what ions would precipitate?
9. The solution of what single ion can be added to solutions of each of the following to distinguish between the pairs? Write equations for the reactions that would occur:
   a. KNO_3 and $Pb(NO_3)_2$.
   b. $Pb(NO_3)_2$ and $Zn(NO_3)_2$.
   c. NaCl and $NaNO_3$.
   d. H_2SO_4 and HCl.
   e. $Ag(NH_3)_2^+ + Cl^-$ and $Ag^+ + NO_3^-$.
   f. K_2CrO_4 and $FeCl_3$.

10. A precipitate is formed when HCl is added to a solution. The precipitate is insoluble in both NH_3 solution and hot water. What ion or ions might be present in the original solution and what ion or ions might be absent?
11. If solutions of both HCl and H_2SO_4 give precipitates with a solution, what ion or ions are almost certainly present and what ion or ions may or may not be present?

4

Analysis of Group II, the Copper and Arsenic Divisions

The ions in Group II all form sulfides that are insoluble in dilute acid solution. This fact is applied in separating Group II from subsequent groups whose sulfides are considerably more soluble and do not precipitate in an acid solution where the hydronium ion concentration is 0.2 to 0.5 molar. The steps in the analysis of Group II are outlined in Chart 3.

Step 6. The HCl and H_2O_2 Treatment. A solution made up to contain only metals of Group II should not contain any nitrate ion. However, in an analysis where both Group I and Group II ions might be present, the solution from Step 1, the precipitation of the silver group, will contain nitrate ions. In acid, nitrate ions oxidize sulfide ions, producing sulfur as follows:

$$8H_3O^+ + 3S^{-2} + 2NO_3^- \rightarrow 2NO\uparrow + 12H_2O + \underline{3S} \text{ (White or yellow).}$$

This sulfur would contaminate sulfides precipitated in Group II and at the same time the precipitating agent, sulfide ions, would be partially if not completely destroyed. The nitrate, if present, is therefore destroyed by boiling the solution from Step 1 with HCl solution. Hydrogen peroxide is added along with the HCl solution in order to convert Sn(II), Sn^{+2}, to Sn(IV), Sn^{+4}. If tin is allowed to remain divalent, SnS will be precipitated in Step 7 and it will not dissolve readily in ammonium sulfide in Step 8. SnS_2, formed from Sn^{+4}, will dissolve readily when treated with ammonium sulfide in Step 8.

Place the solution from Step 1, or ten drops of the solution containing only ions of Group II, in a small beaker. Add 3 drops of 3% H_2O_2 (hydrogen peroxide) solution and boil down to about half the original volume. Then add 3 or 4 drops of concentrated HCl solution

CHART 3

Analysis of Group II, the Copper and Arsenic Divisions

The solution from Step 1 or a solution known to contain only elements of Group II may contain any or all of the following:

Hg^{+2}, Pb^{+2}, Bi^{+3}, Cu^{+2}, Cd^{+2}, $HAsO_2$, $H_2AsO_4^-$, $SbCl_6^{-3}$, Sn^{+2}, $SnCl_6^{-2}$ plus ions from Groups III, IV, and V.

Step 6 Add $\begin{cases} H_2O_2, \\ HCl. \end{cases}$

Boil to small volume.

↓

Hg^{+2}, Pb^{+2}, Bi^{+3}, Cu^{+2}, Cd^{+2}, $H_2AsO_4^-$, $SbCl_6^{-3}$, $SnCl_6^{-2}$ plus ions of the elements of Groups III, IV, and V.

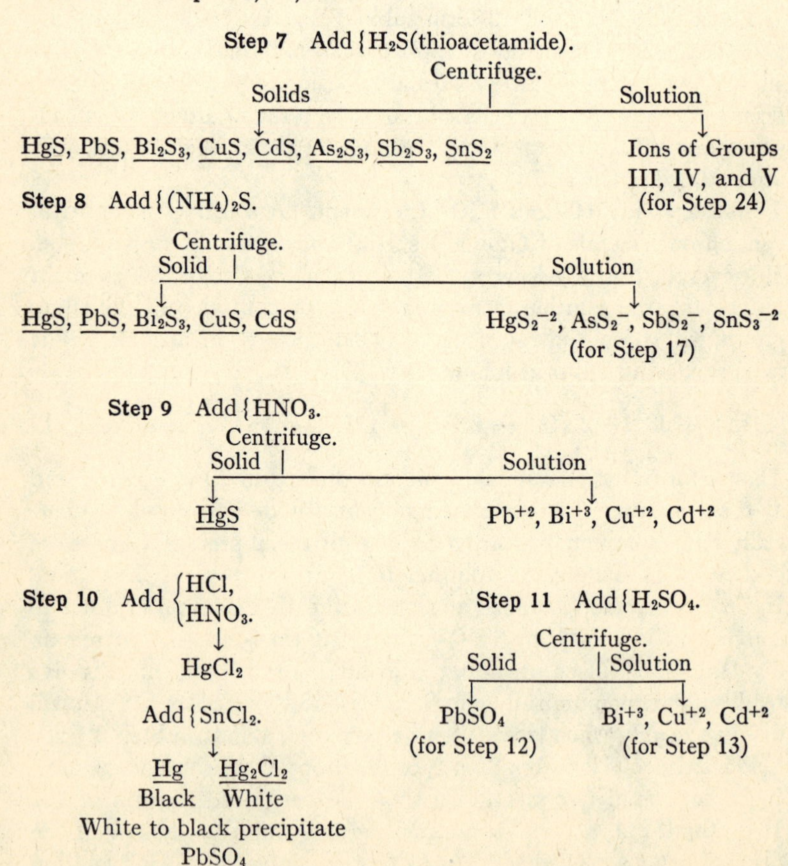

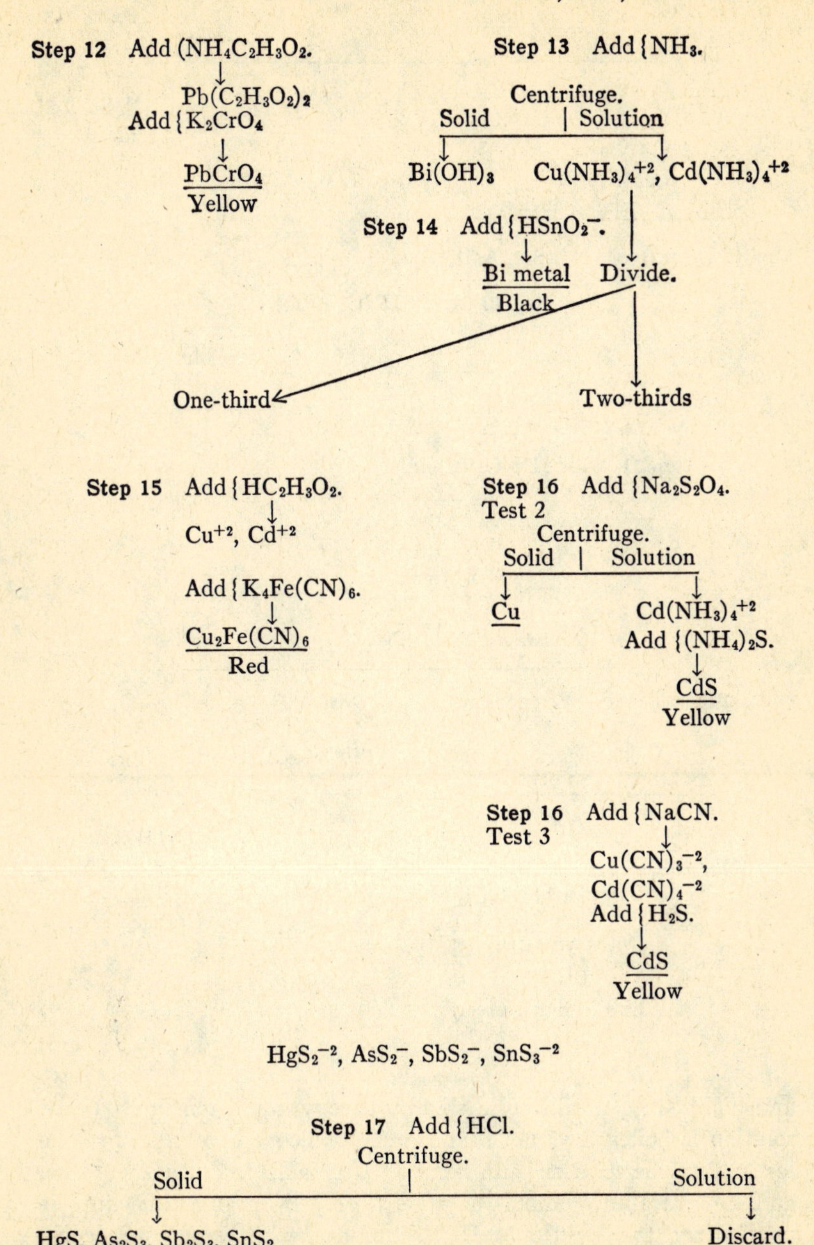
Bi^{+3}, Cu^{+2}, Cd^{+2}
Step 12 Add $NH_4C_2H_3O_2$.
$Pb(C_2H_3O_2)_2$
Add K_2CrO_4
$PbCrO_4$
Yellow
Step 13 Add NH_3.
Centrifuge.
Solid
Solution
$Bi(OH)_3$
$Cu(NH_3)_4^{+2}$, $Cd(NH_3)_4^{+2}$
Step 14 Add $HSnO_2^-$.
Bi metal
Black
Divide.
One-third
Two-thirds
Step 15 Add $HC_2H_3O_2$.
Cu^{+2}, Cd^{+2}
Add $K_4Fe(CN)_6$.
$Cu_2Fe(CN)_6$
Red
Step 16 Add $Na_2S_2O_4$.
Test 2
Centrifuge.
Solid
Solution
Cu
$Cd(NH_3)_4^{+2}$
Add $(NH_4)_2S$.
CdS
Yellow
Step 16 Add NaCN.
Test 3
$Cu(CN)_3^{-2}$,
$Cd(CN)_4^{-2}$
Add H_2S.
CdS
Yellow
HgS_2^{-2}, AsS_2^-, SbS_2^-, SnS_3^{-2}
Step 17 Add HCl.
Centrifuge.
Solid
Solution
HgS, As_2S_3, Sb_2S_3, SnS_2
Discard.

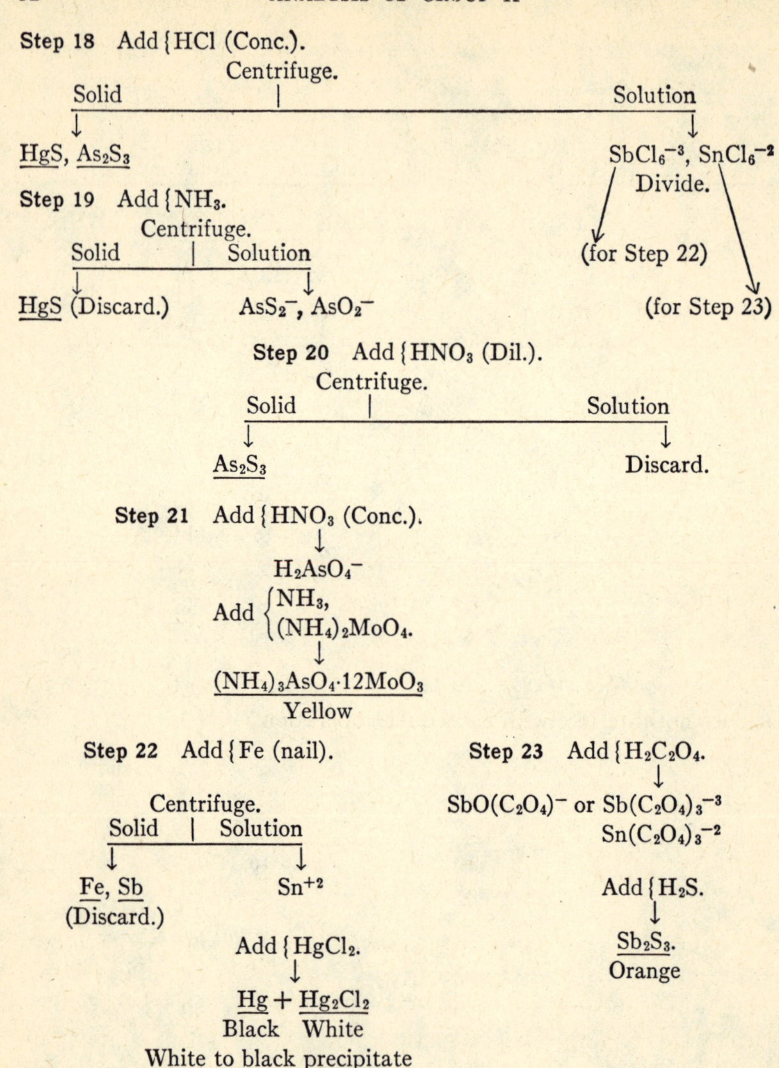

and boil gently, until only a slightly syrupy liquid remains. If the solution is boiled dry, $FeCl_3$ and $SnCl_4$, which are volatile, may be lost. Test the solution with litmus. If it is acid, and it should be, add concentrated ammonia solution a very little at a time until it is alkaline; then add 3 M HCl solution slowly, testing frequently with litmus until the solution is barely acid. Add distilled water until a volume of 1.0 ml is obtained. To this, add 4 drops of 3 M HCl solution. The solution should now be about 0.5 M in HCl and is ready for Step 7.

Heating with HCl not only destroys nitrate ions but also reduces chromates, if present, to Cr^{+3}, to be detected in Group III later. However, chloride ions are not sufficiently strong as a reducing agent to interfere with the oxidation of Sn^{+2} to Sn^{+4} by peroxide ions from hydrogen peroxide.

EQUATIONS FOR REACTIONS IN STEP 6. The main reaction is reduction of the nitrate ion to NO, a gas which evaporates off:

$$8H_3O^+ + 6Cl^- + 2NO_3^- \rightarrow 3Cl_2\uparrow + 2NO\uparrow + 12H_2O.$$

If chromate is present, it is reduced thus:

$$16H_3O^+ + 2CrO_4^{-2} + 6Cl^- \rightarrow 2Cr^{+3} + 3Cl_2\uparrow + 24H_2O.$$

It is essential that tin, if present, all be in the tetravalent state. The oxidation of Sn^{+2} by peroxide is:

$$4H_3O^+ + Sn^{+2} + O_2^{-2} + 6Cl^- \rightarrow SnCl_6^{-2} + 6H_2O.$$

Hydrogen peroxide is slowly decomposed during the boiling so that it does not interfere with subsequent operations:

$$2H_2O_2 \rightarrow 2H_2O + O_2\uparrow.$$

Step 7. Precipitation of Group II Sulfides. The procedure described here employs thioacetamide as the source of sulfide ions. If H_2S is used instead, see Step 7A.

This step must be completed rapidly, if Group III ions may be present. Under no circumstances should the precipitate obtained in this precipitation remain more than a few minutes in contact with a solution which may contain ions of Group III. The sulfides of mercury and bismuth are especially prone to become contaminated with sulfides of Group III by what is known as *postprecipitation.* The sulfides of Group II precipitate normally, but within a few minutes the sulfides of Group III, especially ZnS, if present, begin to grow on the surface of the Group II sulfides. It has been shown that up to 90% of the Zn ions in a solution may postprecipitate on HgS within 10 minutes after the HgS is precipitated, without the solution actually becoming saturated with ZnS. It is therefore necessary to hasten the precipitation and the centrifuging of the sulfides of Group II.

Before anything else is done, prepare a solution for washing the precipitate in order that it will be ready for use. Add 5 drops of 4 M ammonium chloride solution to 1 ml of distilled water in a test tube. Add to this 1 drop of 10% thioacetamide solution. Set this test tube aside and place it in a boiling water bath 1 or 2 minutes before you centrifuge the precipitate of Group II sulfides.

If about one hour of working time is available, place the material from Step 6 in a test tube and add 12 drops of 10% thioacetamide solution. Mix thoroughly. Set the test tube in a beaker of boiling water and keep the water boiling. After 10 minutes add 1.3 ml of distilled water and 3 drops of thioacetamide solution. Mix again and heat in the boiling water for 5 minutes more. Centrifuge at once. The precipitate, if any, is made up of sulfides of such elements of Group II as are present. Decant the solution into a test tube for Step 24 (see Chapter 5), unless Group III, IV, and V ions are known to be absent. In that case discard the solution. The precipitate must be washed. Add to the precipitate about 10 drops of the previously prepared wash solution, mix thoroughly, centrifuge, and add the solution to that saved for Step 24. Wash the precipitate again with another 10 drops of wash solution, centrifuge, and discard this wash solution. Save the precipitate for Step 8.

Place the test tube containing the solution for Step 24 in boiling water for 10 to 15 minutes, transfer the solution to a small beaker, and boil gently over a low flame to a volume of 5 or 6 drops. Heating hydrolyzes thioacetamide in the wash water to H_2S (see p. 65); boiling removes H_2S and some of the excess HCl from the solution. The 5 or 6 drops of solution are now ready for Step 24, the precipitation of Group III, unless interfering anions must be removed.

If the sample being analyzed is a salt or a mixture of salts and if any or all of the borate, fluoride, oxalate, or phosphate anions were found to be present in the sample by doing Steps 65, 66, 67, and 68, then the solution ready for Step 24 must first be treated as described in the section labeled "The Elimination of Interfering Anions," p. 153.

Step 7A. Alternative Method of Precipitating Group II. This method employs H_2S gas as the precipitating agent. This step must be completed as quickly as possible unless Group III, IV, and V ions are known to be absent. The reason for this is given in Step 7.

Connect a glass tube, drawn out and cut off so that the small end will allow H_2S gas to pass through slowly, to the source of H_2S gas.

This source may be a Kipp generator or a large test tube containing an H_2S cartridge and supplied with a one-hole stopper through which a U tube is inserted. Turn on or begin generating H_2S slowly, lowering the small end of the delivery tube below the surface of the solution from Step 6, and rather slowly lower the tip to the bottom of the test tube. Small bubbles should form rather rapidly in the solution. Pass H_2S gas into the solution about 30 seconds. Add 12 drops of distilled water to the solution and pass H_2S through it for 30 seconds more. Centrifuge at once. Decant the solution, boil out the H_2S from it, and save it for Step 24 unless it is known that no ions from Groups III, IV, and V are present. Wash the solid remaining in the test tube twice with 20 drops of distilled water, centrifuge, and discard the washings. The solid is now ready for Step 8.

H_2S gas is exceedingly poisonous. Under no circumstances should one inhale this gas, even in small amounts. Do all work with it under a hood.

If the sample being analyzed is a salt or a mixture of salts and if any or all of the borate, fluoride, oxalate, or phosphate anions were found to be present in the sample by doing Steps 65, 66, 67, and 68, then the solution ready for Step 24 must first be treated as described in the section labeled "The Elimination of Interfering Anions," p. 153.

EQUATIONS FOR REACTIONS IN STEPS 7 AND 7A. Except for the hydrolysis of thioacetamide in Step 7 the equations for reactions in Steps 7 and 7A are the same.

Thioacetamide slowly hydrolyzes with hot water, producing H_2S. This gradually increases the sulfide ion concentration throughout the solution, a process which results in what is known as *homogeneous precipitation*. Precipitates formed by homogeneous precipitation are more pure, of larger crystalline size, and thus more easily separated from solutions by centrifuging than precipitates prepared by adding H_2S gas directly to the solution. The equation is:

$$CH_3CSNH_2 + 2H_2O \rightarrow CH_3COONH_4 + H_2S \uparrow .$$

The H_2S acts as a weak acid with water to produce sulfide ions:

$$H_2S + 2H_2O \rightleftharpoons 2H_3O^+ + S^{-2}.$$

The sulfide ions then react with the ions containing the elements of Group II to form sulfides; the formulas of the precipitates are underlined and the color of the solid is given:

$$Hg^{+2} + S^{-2} \rightarrow \underline{HgS} \text{ (Black; rarely, red).}$$
$$Pb^{+2} + S^{-2} \rightarrow \underline{PbS} \text{ (Black).}$$
$$2Bi^{+3} + 3S^{-2} \rightarrow \underline{Bi_2S_3} \text{ (Brown).}$$
$$Cu^{+2} + S^{-2} \rightarrow \underline{CuS} \text{ (Black).}$$
$$Cd^{+2} + S^{-2} \rightarrow \underline{CdS} \text{ (Yellow).}$$
$$2SbCl_6^{-3} + 3S^{-2} \rightarrow 6Cl^- + \underline{Sb_2S_3} \text{ (Orange).}$$
$$SnCl_6^{-2} + 2S^{-2} \rightarrow 6Cl^- + \underline{SnS_2} \text{ (Yellow).}$$
$$2H_2AsO_4^- + 5S^{-2} + 12H_3O^+ \rightarrow 20H_2O + \underline{2S} + \underline{As_2S_3} \text{ (Yellow).}$$

There is evidence that thioacetamide acts directly on metal ions without first reacting with water to give H_2S. The resulting products are sulfides of metals, whether H_2S is first formed or not.

Step 8. Separation of the Arsenic Division from the Copper Division. The three elements arsenic, antimony, and tin are commonly considered as comprising the arsenic division of Group II. The other elements, all metals, make up the copper division. The separation of the two divisions is based on what is known as the *amphoteric* nature of the oxides and sulfides of the arsenic division elements. The oxides and sulfides of these elements will dissolve in either acid or basic solutions. When they dissolve in acid solutions, positive ions of the elements are formed such as exist in the solid oxides and sulfides — for example, As_2O_3, As_2S_3, SnO_2, SnS_2, Sb_2O_3, and Sb_2S_3. In either basic solutions or solutions of high sulfide-ion concentration, these oxides and sulfides form soluble negative ions such as AsO_2^-, AsS_2^-, SnS_3^{-2}, $Sn(OH)_6^{-2}$, SbS_2^-, and $Sb(OH)_4^-$. Those containing sulfide ions are, of course, formed in strong sulfide solutions, the others in strongly basic solutions. It is obvious at a glance that the oxides and sulfides of these elements are analogous in that they react similarly, oxide ions and hydroxide ions dissolving oxides and sulfide ions dissolving sulfides. The sulfides of the copper division elements do not form complex or negative ions in this way and are not dissolved by strong hydroxide or sulfide solutions. The reagent used here is $(NH_4)_2S$ solution to give a strong sulfide ion concentration. A strong hydroxide ion solution will also extract the arsenic division sulfides away from the copper division sulfides. If a hydroxide solution is used, however, HgS dissolves to some degree, along with the sulfides of the arsenic division. Some books contain procedures which employ a hydroxide solution rather than a sulfide solution.

Because HgS, CuS, and CdS may dissolve slowly in strong sulfide solutions, this step must be completed without delay. Twenty or

25 minutes should be available for work before this step is begun. It is best if this separation is made on freshly precipitated sulfides from Step 7, although this is not always possible.

To the precipitate from Step 7 add 15 drops of $(NH_4)_2S$ solution. Stir steadily for about 60 seconds, centrifuge, and save the solution in a stoppered test tube for Step 17. Add 10 drops of $(NH_4)_2S$ solution to the precipitate and again stir for 60 seconds, centrifuge, and add this solution to that obtained by the first extraction. Now wash the precipitate twice with a hot solution of ammonium nitrate, made by adding 1 drop of 1 M NH_4NO_3 solution to 15 drops of water. Centrifuge each time and discard the washings. The solid contains sulfides of the copper division. Save it for Step 9. This precipitate should be kept wet until Step 9 is begun. Oxygen from the air will slowly convert sulfides to sulfates. Therefore this precipitate should be kept out of contact with air. If starting Step 9 must be delayed until a later laboratory period, boil the dissolved oxygen out of a little water, fill the test tube containing the precipitate with the boiled water, and stopper it. The precipitate is thus kept from drying out and is not contacted by oxygen. If sulfate were formed in the presence of lead ions or PbS, $PbSO_4$ would be produced, which would not dissolve in the nitric acid in Step 9.

EQUATIONS FOR REACTIONS IN STEP 8. All of the reactions in Step 8 are those of formation of soluble negative ions:

$$As_2S_3 + S^{-2} \rightarrow 2AsS_2^-. \quad \text{Metathioarsenite}$$
$$Sb_2S_3 + S^{-2} \rightarrow 2SbS_2^-. \quad \text{Metathioantimonite}$$
$$SnS_2 + S^{-2} \rightarrow SnS_3^{-2}. \quad \text{Trithiostannate(IV)}$$

Step 9. Separation of Mercury from Other Elements of the Copper Division. Advantage is taken of the great insolubility of HgS in water to effect this separation. The other sulfides of the copper division dissolve sufficiently in water so that the sulfide ion concentration of a saturated solution of their sulfides is great enough for 3 M HNO_3 to oxidize the sulfide ion from them to free sulfur. HgS dissolves to such a slight extent ($K_{sp} = 1 \times 10^{-54}$) that 3 M HNO_3 oxidizes the sulfide from it extremely slowly.

The precipitate from Step 8 will be in a test tube. If water was added to the precipitate at the end of Step 8, centrifuge, discard the water, and then add 20 drops, about 1 ml, of 3 M HNO_3 to the solid. Heat by dipping the test tube in boiling water exactly 3 minutes. Longer heating will dissolve some HgS and may convert some sulfide to sulfate, precipitating $PbSO_4$. Centrifuge and decant the liquid into a small beaker or test tube and save for Step 11. Wash the precipitate remaining in the test tube with 10 drops of a mixture of 5 drops of 3 M HNO_3 and 5 drops of distilled water. Centrifuge and discard the wash solution. Save the precipitate in a test tube for Step 10. A black precipitate at this point may be taken as evidence that mercury is probably present in the sample, but it is not conclusive. Some sulfur will be formed as a result of the action of sulfide with nitric acid. The sulfur may surround and entrain some small amount of CuS or PbS and appear to be a black residue like HgS. Mercury must be tested for and its presence confirmed by Step 10.

EQUATIONS FOR REACTIONS IN STEP 9. The sulfides of Pb, Bi, Cu, and Cd all dissolve very slightly in water. In 3 M HNO_3 solution the small amount of sulfide ion in saturated solutions of these sulfides is oxidized by the nitric acid. The dissolving of PbS and Bi_2S_3 can be used to illustrate the production of some sulfide ions in water as indicated by the equations:

$$PbS \rightleftharpoons Pb^{+2} + S^{-2}.$$
$$Bi_2S_3 \rightleftharpoons 2Bi^{+3} + 3S^{-2}.$$

CuS and CdS react in the same way to form ions. The reaction of HNO_3 with sulfide ions can be expressed by the equation:

$$8H_3O^+ + 2NO_3^- + 3S^{-2} \rightarrow \underline{3S} + 2NO\uparrow + 12H_2O.$$

Prolonged heating of nitric acid with sulfides and sulfur may result in the formation of sulfate:

$$8H_3O^+ + 8NO_3^- + 3S^{-2} \rightarrow 3SO_4^{-2} + 8NO\uparrow + 12H_2O.$$
$$2NO_3^- + S \rightarrow SO_4^{-2} + 2NO\uparrow.$$

Sulfate, if formed by either reaction, will precipitate out lead as $PbSO_4$, if lead is present:

$$Pb^{+2} + SO_4^{-2} \rightarrow \underline{PbSO_4}\ (\text{White}).$$

Lead sulfate precipitated thus will not dissolve in nitric acid and will not be found in Step 12.

Step 10. The Test for Mercury. The precipitate will be HgS mixed with sulfur and perhaps small traces of PbS, Bi_2S_3, CuS, CdS, and $PbSO_4$.

Add 6 drops of concentrated HCl and 3 drops of concentrated nitric acid solution to the HgS precipitate in the test tube from Step 9.

Stir and warm in a boiling-water bath about a minute. Add 12 drops of water and boil gently over a small flame for 30 seconds. Centrifuge and decant the clear solution into a test tube. Discard any solid (which is mostly sulfur). Cool the solution and add 3 or 4 drops of 0.2 M stannous chloride ($SnCl_2$) solution. Stir and if a whitish, gray, or black precipitate forms, the presence of mercury is confirmed. If there is no precipitate, mercury is absent. The precipitate, if any, will be Hg (black, in fine droplets), Hg_2Cl_2 (white), or a mixture of the two (gray). This test must not be made while the solution is warm or a gray precipitate of metallic tin may form slowly, due to auto-oxidation, a reaction in which tin is both oxidized and reduced, the equation being:

$$2Sn^{+2} + 6Cl^- \rightarrow SnCl_6^{-2} + \underline{Sn} \text{ (Gray)}.$$

EQUATIONS FOR THE REACTIONS IN STEP 10. The first reaction is that of the oxidation of sulfide by nitrate in the presence of chloride ions to form sulfur and $HgCl_4^{-2}$:

$$Hg^{+2} + 3S^{-2} + 2NO_3^- + 4Cl^- + 8H_3O^+ \rightarrow HgCl_4^{-2} + 2NO\uparrow + 12H_2O + \underline{3S} \text{ (White or yellow)}.$$

Nitric acid cannot oxidize the very few sulfide ions in a saturated solution of HgS without the aid of chloride ions. The latter combine with what few Hg^{+2} ions there are present in the solution to form the very slightly dissociated $HgCl_4^{-2}$ ion. HgS is thus induced to dissolve considerably more in a chloride solution than in water, increasing the sulfide ion concentration to a point where it is oxidized by nitrate ions rather rapidly.

The second reaction is the reduction of mercuric ion to mercury or mercurous chloride by stannous chloride. There are four possible ways this may be expressed by equations, since there are two possible starting compounds containing mercury and two possible products. The four equations are:

$$2Hg^{+2} + 8Cl^- + Sn^{+2} \rightarrow SnCl_6^{-2} + \underline{Hg_2Cl_2} \text{ (White)}.$$
$$Hg^{+2} + 6Cl^- + Sn^{+2} \rightarrow SnCl_6^{-2} + \underline{Hg} \text{ (Black)}.$$
$$2HgCl_4^{-2} + Sn^{+2} \rightarrow SnCl_6^{-2} + \underline{Hg_2Cl_2} \text{ (White)}.$$
$$HgCl_4^{-2} + 2Cl^- + Sn^{+2} \rightarrow SnCl_6^{-2} + \underline{Hg} \text{ (Black)}.$$

Step 11. Separation of Lead from Bismuth, Copper, and Cadmium. The solution from Step 9 may contain ions of the copper division metals, other than mercury, in a solution that is strongly acid with nitric and hydrochloric acids. Lead is separated in this step by converting it to insoluble lead sulfate. The sulfates of the other ions are soluble and are separated by centrifuging.

Add 4 or 5 drops of 9 M H_2SO_4 to the solution from Step 9. Make 9 M sulfuric acid by adding 5 drops of concentrated H_2SO_4, a drop at a time, to 5 drops of water in a small beaker. Stir after each drop of acid is added and keep your eyes protected from spattering. Mix the acid with the solution from Step 9 and set it aside for not less than 10 minutes — and overnight if possible. Gently swirl the solution in the beaker and look closely at the center of the beaker against a black background. If tiny white specks of solid appear to collect in the bottom of the beaker, it is probably $PbSO_4$. There may not be much, but if there is any precipitate at all, centrifuge, decant the solution into a small beaker, and save it for Step 13. Make a wash solution containing 1 drop of concentrated H_2SO_4 in 20 drops of water. Wash the precipitate twice with 10 drops of this solution, centrifuge, and discard both washings. Be careful in decanting not to disturb the precipitate.

EQUATION FOR THE REACTION IN STEP 11. The reaction is the formation of lead sulfate from its ions:

$$Pb^{+2} + SO_4^{-2} \rightarrow \underline{PbSO_4} \text{ (White).}$$

Step 12. The Test for Lead. The $PbSO_4$, if any, is in the bottom of a small test tube after the washing and centrifuging. There should be very little solution in the tube with the precipitate. The test for lead is made by dissolving $PbSO_4$ in 3 M ammonium acetate solution and then precipitating lead chromate from the acetate solution. A very small lead sulfate precipitate may yield a rather large lead chromate precipitate.

Add 10 drops of 3 M $NH_4C_2H_3O_2$ (ammonium acetate) solution to the precipitate. Heat this in boiling water in a beaker for 3 or 4 minutes. If any solid remains undissolved, centrifuge, decant the clear solution, and discard the solid. To the clear solution add 1 drop of 6 M acetic acid and 3 drops of 0.5 M K_2CrO_4 (potassium chromate) solution. If a yellow precipitate forms, it is $PbCrO_4$, which confirms the presence of lead in the sample.

EQUATIONS FOR REACTIONS IN STEP 12. The lead sulfate is dissolved by ammonium acetate solution as a result of the formation of slightly ionized lead acetate (or perhaps some complex containing acetate):

$$PbSO_4 + 2C_2H_3O_2^- \rightarrow Pb(C_2H_3O_2)_2 + SO_4^{-2}.$$

Therefore the concentration of lead ions in solution is reduced below that needed to exceed the solubility product constant for $PbSO_4$.

Lead acetate, however, is dissociated slightly, furnishing a few lead ions to the solution, sufficient to exceed the solubility product constant for $PbCrO_4$ in the chromate solution present. $PbCrO_4$ can precipitate from a solution containing a fairly high concentration of acetate:

$$Pb^{+2} + CrO_4^{-2} \rightarrow \underline{PbCrO_4} \text{ (Yellow)}.$$

This can also be written to indicate the fact that Pb ions can be removed from lead acetate by chromate ions:

$$Pb(C_2H_3O_2)_2 + CrO_4^{-2} \rightarrow 2C_2H_3O_2^- + \underline{PbCrO_4} \text{ (Yellow)}.$$

Step 13. Separation of Bismuth from Copper and Cadmium. The solution from Step 11 is strongly acid with H_2SO_4. It must be neutralized with ammonia and sufficient ammonia in excess must be added to produce soluble ammonia complexes with copper and cadmium ions. Bismuth ions do not form complexes with ammonia but instead form the hydroxide, which precipitates. This separation is an application of the formation of complex ions (see p. 20).

To the solution in the beaker from Step 11, add 1–1 ammonia solution (one part water with one part concentrated ammonium hydroxide) drop by drop, stirring vigorously. Continue adding ammonia solution slowly until the acid is neutralized. Test occasionally with litmus. After the solution is neutral, add 3 drops of concentrated ammonia solution. If copper is present a deep-blue solution will result, although a blue solution at this point may be formed without any copper being present. Further confirmation is necessary. A faintly discernible white precipitate, $Bi(OH)_3$, may be formed. Whether a precipitate is seen or not, centrifuge, and save the solution for Step 15. Wash the precipitate (if any) twice with 10 drops of distilled water. Centrifuge and discard the wash solutions. Save the precipitate of $Bi(OH)_3$ for Step 14.

EQUATIONS FOR REACTIONS IN STEP 13. The addition of a solution of NH_3 to the solution containing H_2SO_4, Bi^{+3}, Cu^{+2}, and Cd^{+2} serves a twofold purpose. The NH_3 neutralizes the sulfuric acid, increasing the hydroxide ion concentration of the solution until bismuth hydroxide precipitates. Neutralization of sulfuric acid may be indicated in two ways. The reaction is between hydronium ions and ammonia or hydroxide ions:

$$H_3O^+ + NH_3 \rightarrow NH_4^+ + H_2O.$$

Ammonia reacts with water to a very slight degree to produce hydroxide ions:

$$NH_3 + H_2O \rightarrow NH_4^+ + OH^-.$$

The hydroxide ions may be said to react with H_3O^+ to form H_2O. In addition, hydroxide ions increase in numbers as ammonia solution is added until the solubility product constant for $Bi(OH)_3$ is exceeded and it precipitates:

$$Bi^{+3} + 3OH^- \rightarrow \underline{Bi(OH)_3} \text{ (White)}.$$

The other purpose of ammonia is to displace water of hydration of copper and cadmium ions, forming ammonia complexes:

$$Cu(H_2O)_4^{+2} + 4NH_3 \rightarrow Cu(NH_3)_4^{+2} + 4H_2O.$$
$$Cd(H_2O)_4^{+2} + 4NH_3 \rightarrow Cd(NH_3)_4^{+2} + 4H_2O.$$

These ammonia complexes are stable in strong ammonia solution. The copper ammonia complex is intensely blue in rather dilute solution and is easily seen in extremely dilute solutions. The cadmium complex is colorless. Nickel also forms an intense-blue complex with ammonia. Although nickel should not be present in the test solution for copper, it may be, so that a further confirmation of the presence of copper in Step 15 is necessary.

When bismuth hydroxide is precipitated, both cadmium and copper hydroxides may precipitate also. However, both are dissolved by an excess of NH_3, thus:

$$Cu(OH)_2 + 4NH_3 \rightarrow Cu(NH_3)_4^{+2} + 2OH^-.$$
$$Cd(OH)_2 + 4NH_3 \rightarrow Cd(NH_3)_4^{+2} + 2OH^-.$$

Step 14. The Test for Bismuth. The precipitate in the test tube, from Step 13, is probably $Bi(OH)_3$. The test for bismuth is carried out by preparing stannite ions in contact with bismuth hydroxide. Stannite ions reduce the bismuth compound to metallic bismuth in a very finely divided state so that it looks velvety black.

Add 4 drops of 6 M NaOH solution to the precipitate. Stir and cool under the cold water faucet a minute or two. Then add 2 drops of stannous chloride solution ($SnCl_2$) and stir with a glass rod. The instantaneous formation of a jet-black precipitate of Bi when the solution is stirred confirms the presence of bismuth. There may be a gray precipitate of tin, formed slowly by auto-oxidation, as indicated by the equation in Step 10, if the solution is not properly cooled.

Heat is generated by NaOH reacting with HCl used in making up the $SnCl_2$ solution.

EQUATIONS FOR REACTIONS IN STEP 14. The first reaction is the formation of stannite ions. In this process $Sn(OH)_2$, a white solid, is first formed. It then is dissolved in an excess of OH^-, forming the stannite ion, $HSnO_2^-$:

$$Sn^{+2} + 2OH^- \rightarrow \underline{Sn(OH)_2} \text{ (White).}$$
$$Sn(OH)_2 + OH^- \rightarrow HSnO_2^- + H_2O.$$

Then stannite reacts immediately with $Bi(OH)_3$ in the strongly basic solution to produce the very black metallic bismuth:

$$3H_2O + 2Bi(OH)_3 + 3HSnO_2^- + 3OH^- \rightarrow 3Sn(OH)_6^{-2} + \underline{2Bi} \text{ (Black).}$$

Step 15. The Test for Copper. Copper ions can be detected in very dilute solutions by forming $Cu_2Fe(CN)_6$, which is red. Whether the solution from Step 13 is blue or not, this test should be made, because a trace of copper may not produce a blue color that can be seen but will still give a good test with ferrocyanide.

Take one-third of the solution from Step 13 for this test and place it in a test tube. Save the other two-thirds for the cadmium test in Step 16.

To the solution for the copper test add 6 M acetic acid ($HC_2H_3O_2$) 2 drops at a time, testing with litmus by removing a part of a drop on a stirring rod and touching it to litmus paper. When the solution tests acid, add two drops of 0.2 M potassium ferrocyanide solution, $K_4Fe(CN)_6$. The proper name for the ferrocyanide ion is the *hexacyanoferrate(II) ion.* Mix well. A brick-red precipitate indicates that copper is present. A reddish cloudiness indicates the presence of small amounts of copper. A greenish precipitate indicates nickel was precipitated with the sulfides of Group II. Green may hide or discolor the red of copper ferrocyanide.

EQUATIONS FOR REACTIONS IN STEP 15. The copper ammonia complex is decomposed by changing NH_3 to NH_4^+, employing acetic acid to furnish the protons needed. With the ammonia concentration greatly reduced, copper ions are released from the blue copper ammonia complex and become available to react with ferrocyanide ions:

$$Cu(NH_3)_4^{+2} + 4HC_2H_3O_2 \rightarrow Cu^{+2} + 4NH_4^+ + 4C_2H_3O_2^-.$$

The copper ions thus freed react with ferrocyanide ions from potassium ferrocyanide:

$$2Cu^{+2} + Fe(CN)_6^{-4} \rightarrow \underline{Cu_2Fe(CN)_6} \text{ (Red)}.$$

Step 16. The Test for Cadmium. There are three possible tests for cadmium. The first is made after the copper test is completed and only if copper is found to be absent. The second, made if copper is found to be present, relies on a powerful reducing agent, $Na_2S_2O_4$, sodium dithionate (named sodium hydrosulfite in commerce), to reduce copper ions to metallic copper. Cadmium ions, if present, are more difficult to reduce than copper ions and remain in solution to react with sulfide ions, giving a yellow precipitate of CdS. The third test is an alternative to the second. It relies on the ability of cyanide ions to reduce Cu^{+2} to Cu^{+} and form a stable complex ion (see p. 23) which does not leave enough Cu^{+2} or Cu^{+} in solution to precipitate CuS or Cu_2S in the presence of sulfide ions. The $Cd(CN)_4^{-2}$ formed is sufficiently unstable to yield enough cadmiun ions to the solution so that when sulfide ions are added the solubility product constant for CdS is exceeded and a yellow precipitate confirms the presence of cadmium. Thus, though there are three different ways of testing for cadmium, all depend on the formation of yellow CdS to confirm the presence of cadmium.

Test 1. If copper is found to be absent, add 2 drops of ammonium sulfide solution to the remaining two-thirds of the solution from Step 15. If cadmium is present, a yellow precipitate of CdS will form at once.

Test 2. If copper is found to be present, add an amount of solid sodium dithionate equal in size to a BB shot to the remaining two-thirds of the solution from Step 15. Stir and warm the test tube in a beaker of boiling water for 2 minutes. Copper separates from the solution as a black, powdery solid. Centrifuge and decant the clear liquid, which must be colorless at this point. If the solution is blue, add more sodium dithionate and heat again. Discard the precipitate; it is copper metal. Place the clear, colorless solution in a test tube and add 2 drops of ammonium sulfide solution. A yellow precipitate of CdS confirms the presence of cadmium.

Test 3. If copper is found to be present, test the remaining two-thirds of the solution from Step 15 with litmus. If the solution is not basic, add ammonia solution until it is. Then add 0.2 M KCN solution drop by drop with stirring until all blue color disappears.

The copper ammonia complex is converted to the cuprous cyanide complex $Cu(CN)_3^{-2}$. The cadmium ammonia complex is also converted to the cyanide complex $Cd(CN)_4^{-2}$. Add 2 drops of ammonium sulfide solution. If a yellow precipitate forms, the presence of cadmium is confirmed. After this test is completed, the solution should be washed into the sink with plenty of water. Cyanides must never be used carelessly. If added to acid solutions, cyanide ions form HCN, a very poisonous gas. It is recommended that cyanide solution be added by the instructor, or by the stock room man, who is also well acquainted with safety procedures.

EQUATIONS FOR REACTIONS IN STEP 16. For Test 1, the only reaction is between cadmium and sulfide ions:

$$Cd^{+2} + S^{-2} \rightarrow \underline{CdS} \text{ (Yellow)}.$$

For Test 2, the reduction of copper ions by the dithionate ion is illustrated by the equation:

$$Cu(NH_3)_4^{+2} + S_2O_4^{-2} + 2H_2O \rightarrow 4NH_4^+ + 2SO_3^{-2} + \underline{Cu} \text{ (Black)}.$$

Then CdS is precipitated by the reaction between cadmium and sulfide ions as in Test 1.

For Test 3, cyanide converts both the copper and cadmium ammonia complexes to the cyanide complexes, reducing Cu(II) to Cu(I), while cyanide is oxidized to CNO^-, the cyanate:

$$2Cu(NH_3)_4^{+2} + 7CN^- + 2OH^- \rightarrow 2Cu(CN)_3^{-2} + 8NH_3 + CNO^- + H_2O.$$

$$Cd(NH_3)_4^{+2} + 4CN^- \rightarrow Cd(CN)_4^{-2} + 4NH_3.$$

The cyanide complex dissociates to give a few cadmium ions:

$$Cd(CN)_4^{-2} \rightleftharpoons Cd^{+2} + 4CN^-.$$

Since the Cd ion concentration from this reaction exceeds the solubility product constant of CdS, if sulfide ions are added, CdS precipitates. The equation for the reaction is the same as that for Test 1. See p. 23 for calculations concerning this test.

Step 17. Precipitation of the Sulfides of the Arsenic Division. The solution from Step 8 may contain AsS_2^-, SbS_2^-, SnS_3^{-2}, and perhaps a small amount of HgS_2^{-2}, along with a large amount of ammonium sulfide, all in solution. HCl added will increase the concentration of hydronium ions; these combine with and reduce the sulfide ion concentration to a point where the sulfide complexes are converted to the sulfides, which precipitate.

The solution from Step 8 should be in a test tube. Add to it 6 M HCl drop by drop, stirring with a glass rod and testing with litmus frequently by removing a small drop on the rod to touch to litmus paper. As soon as the solution tests acid, all of the sulfides should be precipitated. (If there is no precipitate, all of the elements of the arsenic division are absent; thus Steps 18 through 23 can be omitted.) Centrifuge, decant, and discard the solution. Wash the precipitate twice with hot distilled water, centrifuging and discarding the wash water. The solid is ready for Step 18.

EQUATIONS FOR REACTIONS IN STEP 17. In all cases the sulfide complexes are converted to the sulfides by removing the excess of sulfide ions from solution in the form of H_2S gas, which escapes from the solution.

$$2AsS_2^- + 2H_3O^+ \rightarrow H_2S\uparrow + 2H_2O + \underline{As_2S_3} \text{ (Yellow).}$$
$$2SbS_2^- + 2H_3O^+ \rightarrow H_2S\uparrow + 2H_2O + \underline{Sb_2S_3} \text{ (Orange).}$$
$$SnS_3^{-2} + 2H_3O^+ \rightarrow H_2S\uparrow + 2H_2O + \underline{SnS_2} \text{ (Yellow).}$$

If mercury is present, the following reaction may occur:

$$HgS_2^{-2} + 2H_3O^+ \rightarrow H_2S\uparrow + 2H_2O + \underline{HgS} \text{ (Black).}$$

Step 18. Separation of Arsenic and Mercury from Antimony and Tin. The sulfides of antimony and tin are very insoluble in water, but they are much more soluble in water than As_2S_3 or HgS (if the latter is present). Consequently the sulfides of antimony and tin can be dissolved by concentrated HCl, leaving As_2S_3 or HgS undissolved.

The precipitate from Step 17 is in a small test tube. Add 15 drops of concentrated HCl solution. Mix well with a glass rod, while heating in boiling water in a beaker. Heat and stir 2 minutes. Immediately centrifuge, decant the clear liquid into a test tube, and save it for Steps 22 and 23. Add 10 drops of 6 M HCl solution to the precipitate. Stir thoroughly and centrifuge. Add the washing solution to that saved for Steps 22 and 23. Wash the precipitate again with 10 drops of 6 M HCl. Centrifuge and discard this solution. Save the precipitate for Step 19. If the precipitate is very small in quantity and white in color, none of the arsenic division elements are present and Steps 19 through 23 may be omitted.

EQUATIONS FOR REACTIONS IN STEP 18. The sulfides of antimony and tin dissolve in HCl, forming complex chlorides and H_2S:

$$Sb_2S_3 + 6H_3O^+ + 12Cl^- \rightarrow 2SbCl_6^{-3} + 3H_2S \uparrow + 6H_2O.$$
$$SnS_2 + 4H_3O^+ + 6Cl^- \rightarrow SnCl_6^{-2} + 2H_2S \uparrow + 4H_2O.$$

If water instead of 6 M HCl is used to wash the antimony and tin solution away from the HgS or As_2S_3, $SbCl_6^{-3}$ might react with water (hydrolyze) to give a white solid, SbOCl, which would contaminate the precipitate and cause a loss of some or all of the antimony. The equation for this reaction is:

$$SbCl_6^{-3} + 3H_2O \rightarrow 5Cl^- + 2H_3O^+ + \underline{SbOCl}\ \text{(White)}.$$

Step 19. Separation of Mercury from Arsenic. While HgS should not be mixed with As_2S_3 at this point, it more often than not is and should be separated. HgS is usually black. If the precipitate of As_2S_3 is bright yellow, this step may be omitted. The principle of this separation is the fact that HgS is not soluble in ammonia solution whereas As_2S_3 is very soluble.

To the precipitate from Step 18 add 10 drops of water. Mix well and centrifuge. Discard the water. Add 10 drops of concentrated ammonia solution (ammonium hydroxide) to the solid and stir with a glass rod for 30 seconds. Centrifuge and decant the solution into a small beaker for Step 20. Discard the solid, which will be HgS, sulfur, and other accumulated solids.

EQUATION FOR THE REACTION IN STEP 19. Arsenious sulfide is soluble in the high hydroxide ion concentration of a strong ammonia solution, forming negative ions:

$$2As_2S_3 + 4OH^- \rightarrow 3AsS_2^- + AsO_2^- + 2H_2O.$$

Step 20. The Reprecipitation of Arsenic Sulfide. The solution from Step 19 is strongly alkaline with ammonia. Nitric acid is added to neutralize the ammonia because chloride ions from HCl would interfere with the reaction in Step 21. Otherwise HCl would do just as well to reprecipitate As_2S_3.

Add 3 M HNO_3 1 or 2 drops at a time, stirring gently after each addition, until the solution tests acid to litmus. Add 3 or 4 drops in excess. A yellow precipitate of As_2S_3 will form if arsenic is present, but this is not proof of the presence of arsenic. There may be some whitish sulfur formed by the action of nitric acid on sulfide, and some traces of antimony or tin may give a small amount of color to this sulfur. Centrifuge and discard the solution. Save the solid for Step 21.

EQUATION FOR THE REACTION IN STEP 20. Arsenite and thioarsenite are converted to arsenic sulfide, essentially the reverse of the equation in Step 19:

$$3AsS_2^- + AsO_2^- + 4H_3O^+ \rightarrow 6H_2O + \underline{2As_2S_3} \text{ (Yellow)}.$$

Step 21. The Test for Arsenic. Arsenic sulfide from Step 20 is dissolved in nitric acid, producing arsenate ions. Then ammonium molybdate is added to precipitate a characteristic yellow ammonium molybdoarsenate.

Add 5 drops of concentrated HNO_3 to the precipitate from Step 20. Heat the test tube in boiling water with occasional stirring for 10 minutes. A considerable amount of sulfur may form by interaction of nitric acid and sulfide. From a medicine dropper allow concentrated ammonia solution to trickle slowly down the side of the test tube, shaking after each small addition of ammonia. Test with litmus occasionally and continue to add ammonia until the solution is alkaline. Centrifuge to separate sulfur from the solution. Discard the solid. To the solution add 4 drops of ammonium molybdate, $(NH_4)_2MoO_4$, solution and concentrated nitric acid until at least 5 drops more than enough to neutralize all ammonia have been added. Place the test tube in a beaker of boiling water, stir occasionally, and heat for 5 or 10 minutes. If a yellow precipitate was obtained in Step 20 and no yellow precipitate forms at this point, add 2 more drops of ammonium molybdate solution and 10 drops of concentrated nitric acid. Mix well and heat 5 minutes more. A yellow precipitate is confirmation of the presence of arsenic. The precipitate is not always easily seen. Centrifuging concentrates it in the bottom of the tube, where it is more easily observed.

EQUATIONS FOR REACTIONS IN STEP 21. The first reaction is between nitric acid and arsenic sulfide to produce the acid arsenate ion and sulfur and nitric oxide. There are several reactions involved. Arsenic(III) is oxidized to arsenic(V), sulfide is oxidized to free sulfur, and nitrate is reduced to nitric oxide, NO. The several equations are given, then added together to give the total equation.

The separate equations are:

1. $3S^{-2} + 2NO_3^- + 8H_3O^+ \rightarrow 2NO\uparrow + 12H_2O + \underline{3S}$ (White or yellow).
2. $3As^{+3} + 2NO_3^- + 18H_2O \rightarrow 3H_2AsO_4^- + 2NO\uparrow + 10H_3O^+$.

The combined equation is 3 times Equation 1 plus 2 times Equation 2 added together:

$$3As_2S_3 + 10NO_3^- + 4H_3O^+ \rightarrow 6H_2AsO_4^- + 10NO\uparrow + \underline{9S} \text{ (White or yellow)}.$$

The final test for arsenic is the reaction of arsenate with molybdate in the presence of an excess of ammonium ions. This reaction occurs only in a strongly acid solution; hydronium ions are needed in large numbers to complete the reaction:

$$H_2AsO_4^- + 3NH_4^+ + 12MoO_4^{-2} + 22H_3O^+ \rightarrow 34H_2O + \underline{(NH_4)_3AsO_4 \cdot 12MoO_3} \text{ (Yellow)}.$$

Step 22. The Test for Tin. The solution from Step 18 may contain chloro complexes of tin and antimony. The solution is divided into equal parts, one for testing for tin and one for testing for antimony. The test for tin is made after antimony is removed from the solution by displacement by iron. Aluminum may also be used, but tin and antimony are reduced to the metal by aluminum; so all the aluminum and the precipitated tin must be dissolved completely, a tedious process. Iron in the form of short pieces of wire or tiny nails known as "brads" reduces antimony to the metal and tin to the Sn(II) state without any metallic tin being formed. The mixture is centrifuged and the solid antimony and iron are discarded. The remaining solution can be tested for tin by the same test as that used for mercury in Step 10.

Boil half of the solution from Step 18 for a few seconds to drive off any H_2S; but do not boil dry. Add a small piece of clean, shiny iron wire or a small brad; there must be *no* rust on the iron. Add 10 drops of distilled water. Heat the solution with the tube in boiling water for *exactly* 5 minutes. Centrifuge and decant the solution into a test tube. Discard the nail and solid antimony, if any. To the clear solution add a drop of saturated mercuric chloride ($HgCl_2$) solution. If a white or gray or black precipitate forms, the presence of tin is confirmed.

EQUATIONS FOR REACTIONS IN STEP 22. The first reactions are between iron and Sn(IV) and Sb(III), forming Sn(II) and Sb:

$$SnCl_6^{-2} + Fe \rightarrow Fe^{+2} + Sn^{+2} + 6Cl^-.$$
$$2SbCl_6^{-3} + 3Fe \rightarrow 3Fe^{+2} + 12Cl^- + \underline{2Sb} \text{ (Black)}.$$

The next reaction is the same as that applied in the test for mercury in Step 10:

$$Sn^{+2} + 2HgCl_2 + 4Cl^- \rightarrow SnCl_6^{-2} + \underline{Hg_2Cl_2} \text{ (White)}.$$

In an excess of Sn^{+2}:

$$Sn^{+2} + Hg_2Cl_2 + 4Cl^- \rightarrow SnCl_6^{-2} + \underline{2Hg} \text{ (Black)}.$$

The color of the final precipitate in the test for tin depends on the relative amounts of Sn^{+2} and $HgCl_2$. A large excess of tin yields a black precipitate. An excess of mercuric chloride yields a white precipitate. All shades of gray are possible with different relative amounts of the two reactants.

Step 23. The Test for Antimony. The principle applied here is that tin as Sn(IV) will form an extremely stable complex ion with oxalate, leaving antimony ions in sufficient quantity so that the solubility product constant for antimony sulfide is exceeded when sulfide is added.

Boil the other half of the solution from Step 18 in a small beaker for a few seconds to remove H_2S. Add 0.5 g of solid oxalic acid and 5 ml of distilled water. Mix with a stirring rod until the oxalic acid is dissolved. Add 5 drops of thioacetamide solution. Warm in a boiling-water bath for 2 minutes. If a red-orange precipitate forms, the presence of antimony (as Sb_2S_3) is confirmed. If a brownish precipitate forms after 4 or 5 minutes, this is probably SnS_2, which may form slowly if large amounts of tin are present.

Step 23A. Alternative Test for Antimony. This is known as the "coin test." To make this test, boil the solution from Step 18 to remove H_2S. Add 5 drops of distilled water, and mix well. Place a large drop of this solution on a silver coin with a medicine dropper. Place a piece of metallic tin in contact with the solution and the coin simultaneously. If a very black deposit of Sb forms quickly on the coin under the solution, the presence of antimony is confirmed.

EQUATIONS FOR REACTIONS IN STEPS 23 AND 23A. The reaction of oxalate with tin and antimony ions forms complexes that differ in stability. The tin complex is much more stable than the antimony complex:

$$SnCl_6^{-2} + 3C_2O_4^{-2} \rightarrow Sn(C_2O_4^{-2})_3^{-2} + 6Cl^-.$$

$$SbCl_6^{-3} + C_2O_4^{-2} + 3H_2O \rightarrow SbO(C_2O_4^{-2})^- + 6Cl^- + 2H_3O^+.$$

$SbO(C_2O_4^{-2})^-$, the antimonyl oxalate complex ion, dissociates to yield SbO^+ in sufficient concentration so that if sulfide and hydronium ions are added the solubility product constant for Sb_2S_3 is exceeded and precipitation occurs. The equation for formation of SbO^+ is:

$$SbO(C_2O_4^{-2})^- \rightleftharpoons SbO^+ + C_2O_4^{-2}.$$

Sulfide ions result when thioacetamide hydrolyzes to give H_2S (see p. 65). $(NH_4)_2S$ may be added to furnish sulfide ions, but it should not be used because H_2S gas forms in such large amounts that it bubbles out. H_2S is a very poisonous gas, which must not be inhaled. Sulfide from H_2S reacts with antimonyl ions to produce reddish-orange Sb_2S_3:

$$2SbO^+ + 3S^{-2} + 4H_3O^+ \rightarrow 6H_2O + \underline{Sb_2S_3} \text{ (Reddish-orange).}$$

These last two equations can be added together to give:

$$4H_3O^+ + 2SbO(C_2O_4^{-2})^- + 3S^{-2} \rightarrow C_2O_4^{-2} + 6H_2O + \underline{Sb_2S_3} \text{ (Reddish-orange).}$$

In the alternate test for antimony the reaction is one of displacement or replacement. Metallic tin displaces antimony from solution by forming an electrolytic couple with the silver from the coin:

$$3Sn + 2SbCl_6^{-3} + 6Cl^- \rightarrow 3SnCl_6^{-2} + \underline{2Sb} \text{ (Black).}$$

Review Questions and Problems

1. Write out Chart 3 from memory.
2. From Chart 3 write equations for all reactions in the steps of the analysis of Group II.
3. What principle is applied in each separation accomplished in the steps shown in Chart 3? For example, the arsenic group is separated from the copper group by utilizing the amphoteric nature of the arsenic group elements, forming soluble thio (sulfur) anions that are extracted from the copper group sulfides, which do not form thio anions.
4. Write the formula of a single reagent that will:
   a. Give a precipitate with Sn^{+2} but not with $SnCl_6^{-2}$.
   b. Dissolve SnS_2 but not As_2S_3.
   c. Give a precipitate with Pb^{+2} but not with Bi^{+3}.
   d. Give a precipitate with $(NH_4)_2S$ but not with HCl.
   e. Give a precipitate with $HgCl_2$ but not with Sn^{+2}.
   f. Give a precipitate with H_2SO_4 solution but not with HNO_3 solution.
   g. Oxidize Sn^{+2}.
   h. Dissolve $Cd(OH)_2$ but not $Bi(OH)_3$.
   i. Distinguish between SnS_2 and As_2S_5.
   j. Distinguish between Cd^{+2} and Pb^{+2}.
   k. Distinguish between CdS and As_2S_3.
   l. Dissolve $PbSO_4$ readily.
5. A colorless solution containing elements of Group II, on treatment with H_2S, yielded a yellow precipitate. This precipitate was completely soluble in ammonium sulfide solution. What elements might be present?

6. A greenish-colored Group II solution gave a black precipitate with H_2S (Step 7), which was extracted with $(NH_4)_2S$ (Step 8), and the remaining solid was found to be soluble in 3 M HNO_3 (Step 9). The $(NH_4)_2S$ extract produced an orange precipitate when acidified (Step 17).
   a. What elements are probably present?
   b. What elements are probably absent?
   c. What elements are *possibly* present — that is, no evidence for being absent or for being present is given?
7. A colorless Group II solution gave a yellow sulfide precipitate (Step 7), which was not all dissolved by ammonium sulfide (Step 8); the residue from Step 8 was soluble in 3 M HNO_3 (Step 9), gave no precipitate with sulfate (Step 11), and with ammonia gave a white precipitate which dissolved to give a colorless solution as more ammonia was added in Step 13. The solution from Step 8 gave (Step 17) a yellow precipitate with HCl which dissolved in concentrated HCl in Step 18.
   a. What elements are definitely absent?
   b. What elements are probably absent?
   c. What elements would one expect to be present from the evidence given?
8. If separate solutions contain the following pairs of ions, how are the elements separated in the Group II procedure? Check your answer with Chart 3.

Pb^{+2} and Sb^{+3}
Hg^{+2} and Cu^{+2}
Bi^{+3} and Sn^{+4}
As^{+3} and Sb^{+3}
Cd^{+2} and Cu^{+2}

5

Analysis of Group III, the Nickel and Aluminum Divisions

The ions of the metals of Group III form either sulfides or hydroxides (both insoluble) in a solution made slightly alkaline with ammonia to which sulfide ions are added. NH_4Cl is added to buffer (see p. 36) the solution against an excessively high hydroxide ion concentration from ammonia alone, which might precipitate hydroxides of elements in subsequent groups.

Group III consists of two divisions. One is the nickel division, containing elements whose hydroxides are not amphoteric. The other is the aluminum division, containing elements whose hydroxides are amphoteric and are therefore soluble in a high hydroxide ion concentration. As in the Group II analysis, the amphoteric elements are extracted away from sulfides of the nonamphoteric elements by forming negative ions which are soluble in a strongly basic solution. The steps in the analysis of Group III are shown in Chart 4.

Ions of the subsequent groups — Ba^{+2}, Sr^{+2}, Ca^{+2}, Mg^{+2}, K^+, and Na^+ — form soluble hydroxides, except for Mg^{+2}, which forms a hydroxide so strongly basic that the hydroxide ion concentration needed to exceed its solubility product constant is never attained in the presence of NH_4^+ acting as a buffer.

The sample to be analyzed may contain only ions of the Group III elements or any or all of the ions of Groups III, IV, and V. If the sample is the solution from Step 7, it may therefore contain Mn^{+2}, Fe^{+3}, Co^{+2}, Ni^{+2}, Al^{+3}, Cr^{+3}, Zn^{+2}, and ions of Groups IV and V, plus considerable amounts of HCl not completely removed by the boiling treatment recommended in Step 7 or Step 7A.

Procedures from this point on may vary greatly. Some chemists prefer to remove manganese immediately as MnO_2. Others prefer to

CHART 4

Analysis of Group III, the Nickel and Aluminum Divisions

The solution from Step 7 or a solution known to contain only elements of Group III may contain any or all of the following:

Al^{+3}, Cr^{+3}, Fe^{+3}, Co^{+2}, Ni^{+2}, Mn^{+2}, and Zn^{+2}.

The solution from Step 7 may also contain ions of Groups IV and V.

Step 24 Add { NH_3, NH_4Cl, $(NH_4)_2S$.

Centrifuge.

Solid ↓ $\underline{MnS}$, $\underline{Fe(OH)_3}$, $\underline{FeS}$, $\underline{CoS}$, $\underline{NiS}$, $\underline{Al(OH)_3}$, $\underline{Cr(OH)_3}$, $\underline{ZnS}$

Solution ↓ Ions of Groups IV and V (for Step 40)

Step 25 Add { HCl, HNO_3.

↓

Mn^{+2}, Fe^{+3}, Co^{+2}, Ni^{+2}, Al^{+3}, Cr^{+3}, Zn^{+2}

Step 26 Add { NaOH, H_2O_2.

Centrifuge.

Solid ↓ $\underline{MnO_2}$, $\underline{Fe(OH)_3}$, $\underline{Co(OH)_3}$, $\underline{Ni(OH)_2}$

Solution ↓ $Al(OH)_4^-$, CrO_4^{-2}, $Zn(OH)_4^{-2}$

Add { HNO_3.

↓

Al^{+3}, $Cr_2O_7^{-2}$, Zn^{+2} (for Step 35)

Step 27 Add { HNO_3, H_2O_2 or Na_2O_2.

↓

Mn^{+2}, Fe^{+3}, Ni^{+2}, Co^{+2}

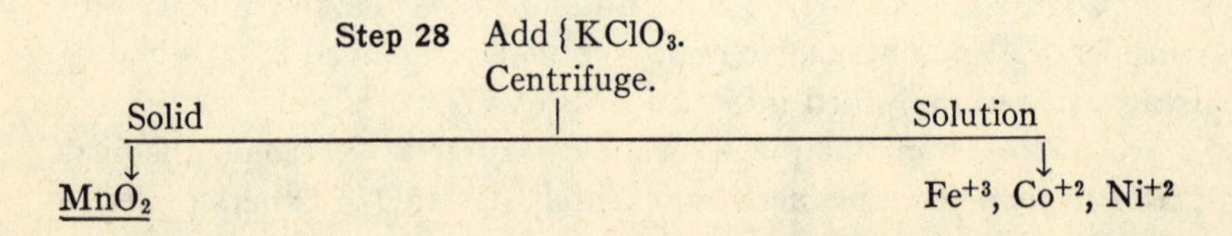

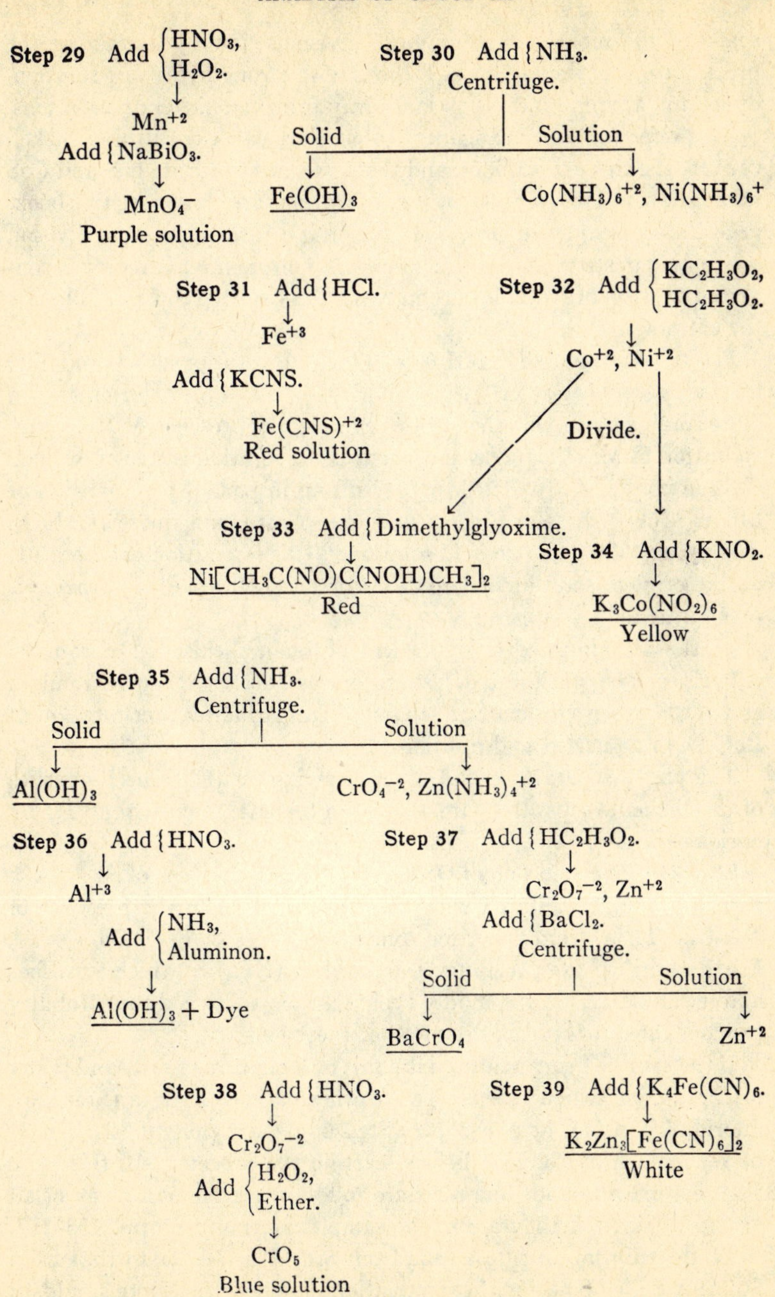
Step 29 Add {HNO_3, H_2O_2.
Mn^{+2}
Add {$NaBiO_3$.
MnO_4^-
Purple solution
Step 30 Add {NH_3.
Centrifuge.
Solid
Solution
$Fe(OH)_3$
$Co(NH_3)_6^{+2}$, $Ni(NH_3)_6^{+}$
Step 31 Add {HCl.
Fe^{+3}
Add {KCNS.
$Fe(CNS)^{+2}$
Red solution
Step 32 Add {$KC_2H_3O_2$, $HC_2H_3O_2$.
Co^{+2}, Ni^{+2}
Divide.
Step 33 Add {Dimethylglyoxime.
$Ni[CH_3C(NO)C(NOH)CH_3]_2$
Red
Step 34 Add {KNO_2.
$K_3Co(NO_2)_6$
Yellow
Step 35 Add {NH_3.
Centrifuge.
Solid
Solution
$Al(OH)_3$
CrO_4^{-2}, $Zn(NH_3)_4^{+2}$
Step 36 Add {HNO_3.
Al^{+3}
Add {NH_3, Aluminon.
$Al(OH)_3$ + Dye
Step 37 Add {$HC_2H_3O_2$.
$Cr_2O_7^{-2}$, Zn^{+2}
Add {$BaCl_2$.
Centrifuge.
Solid
Solution
$BaCrO_4$
Zn^{+2}
Step 38 Add {HNO_3.
$Cr_2O_7^{-2}$
Add {H_2O_2, Ether.
CrO_5
Blue solution
Step 39 Add {$K_4Fe(CN)_6$.
$K_2Zn_3[Fe(CN)_6]_2$
White

separate the nickel and aluminum divisions first by extracting the precipitate containing Group III ions with strong hydroxide solution, dissolving aluminum division ions away from the nickel division ions. Some prefer to allow manganese to go along with cobalt and nickel, convert the three to sulfides, and then extract MnS from NiS and CoS with HCl solution. The following separations are chosen for this book.

Of the hydroxides of Group III ions, those of the aluminum division are soluble in strong hydroxide solution. Chromium hydroxide is converted to the more soluble chromate (Cr^{+3} is oxidized to CrO_4^{-2}) before the extraction.

$Fe(OH)_3$ is insoluble in NH_4Cl and NH_4OH solutions, Co^{+2} and Ni^{+2} form very soluble $Co(NH_3)_6^{+2}$ and $Ni(NH_3)_6^{+2}$, and Mn^{+2} forms a much more soluble hydroxide than does Fe^{+3}. In the presence of NH_4^+ precipitation of $Mn(OH)_2$ does not occur if the solution is free of oxygen. Oxygen converts Mn^{+2} to Mn^{+4}, resulting in partial precipitation of Mn as MnO_2. A safer procedure is to separate manganese as MnO_2, employing chlorate in an acid solution. This separation can be made on the original solution of Group III or just before Fe^{+3} hydroxide is precipitated.

In the aluminum division, where chromium exists as chromate, $Al(OH)_3$ can be precipitated, leaving zinc as $Zn(NH_3)_4^{+2}$ and chromium as CrO_4^{-2}; then chromate is separated from zinc by precipitation of $BaCrO_4$ in an acetic acid solution.

The above separations and others, relying on the same properties of the elements separated, are applied in the analysis of Group III by various authors.

Step 24. Precipitation of Group III. The hydroxides of Al^{+3}, Cr^{+3}, and Fe^{+3} are precipitated with ammonia solution in the presence of NH_4Cl. Then, from the same solution, sulfides of Fe^{+2}, Co^{+2}, Ni^{+2}, Mn^{+2}, and Zn^{+2} are precipitated by adding $(NH_4)_2S$, which furnishes ample sulfide ions in the mildly basic solution to exceed the solubility product constants of sulfides of all of these ions.

If the sample being analyzed is a salt or a mixture of salts and if any of the anions which interfere with other tests — oxalate (Step 66), fluoride (Step 67), or phosphate (Step 68) — are present, they must be removed from solution before starting the procedure of this step. The reasons why and the procedure to follow for removing any or all of these anions, if they are in the sample, are given on pp. 153–157.

To the solution in a test tube from Step 7 or 7A, or to that from Step 81, 82, or 83 after the elimination of interfering anions, add 10 drops of 4 M NH_4Cl solution to provide NH_4^+ as a buffer, and heat to near boiling. Add concentrated NH_3 solution drop by drop, stirring and testing with litmus after each drop is added. After the solution is no longer acid, add 2 drops of concentrated NH_3 solution. If no precipitate forms at this point, Al^{+3}, Cr^{+3}, and Fe^{+3} are absent from the sample. If a precipitate forms, centrifuge and observe the color. $Al(OH)_3$ is white, $Cr(OH)_3$ is green, and $Fe(OH)_3$ is reddish-brown.

Add 5 drops of $(NH_4)_2S$ solution, stir, centrifuge, and allow one more drop of $(NH_4)_2S$ solution to trickle down the side of the test tube. If more precipitate forms in the clear solution above the precipitate, the 5 drops were not sufficient to precipitate all of the ions. Stir, centrifuge, and repeat the procedure, adding a drop of $(NH_4)_2S$ solution each time until no more precipitate is formed. Then decant the clear solution with a medicine dropper with the tip drawn out to a rather fine capillary. Draw the solution from the top of the liquid layer to avoid stirring up the precipitate. If the solution is to be analyzed for Groups IV and V, place it in a small beaker and add 10 drops of concentrated acetic acid (glacial), boil gently to half the original volume to remove all sulfide, transfer the solution to a test tube, stopper the tube, and save it for Step 40, the precipitation of Group IV. If ions of Groups IV and V are known to be absent, discard the solution.

Wash the precipitate remaining in the tube twice, each time using half of a solution made up of 20 drops (1 ml) of distilled water, 2 drops of concentrated ammonia solution, 2 drops of 4 M NH_4Cl solution, and 1 drop of $(NH_4)_2S$ solution. Discard the washings.

The solid may contain $Al(OH)_3$, $Cr(OH)_3$, $Fe(OH)_3$, FeS, CoS, NiS, MnS, and ZnS, and will be in the bottom of the centrifuge tube. It is ready for Step 25.

Equations for Reactions in Step 24. The first reaction is neutralization of the excess HCl in the solution remaining from Step 7 or 7A, by NH_3:

$$NH_3 + H_3O^+ \rightarrow NH_4^+ + H_2O.$$

The NH_4^+ formed augments that from the NH_4Cl added to furnish NH_4^+ as a buffer (see p. 36). If a large excess of NH_3 were added, it would react with water to give a high concentration of hydroxide ion:

$$NH_3 + H_2O \rightleftharpoons NH_4^+ + OH^-.$$

The high concentration of ammonium ions prevents this reaction from continuing further to the right than is necessary to furnish hydroxide ions in sufficient concentration to exceed the solubility product constant for $Al(OH)_3$, $Cr(OH)_3$, and $Fe(OH)_3$. If the reaction indicated by the equation above were allowed to proceed toward the right unchecked by a high concentration of NH_4^+, the solubility product constant for $Mg(OH)_2$ might be exceeded.

The addition of $(NH_4)_2S$ solution furnishes sulfide ions:

$$(NH_4)_2S \rightarrow 2NH_4^+ + S^{-2},$$

which in turn combine with the ions of elements in the nickel division to form sulfides:

$$Fe^{+2} + S^{-2} \rightarrow \underline{FeS} \text{ (Black).}$$
$$Co^{+2} + S^{-2} \rightarrow \underline{CoS} \text{ (Black).}$$
$$Ni^{+2} + S^{-2} \rightarrow \underline{NiS} \text{ (Black).}$$
$$Mn^{+2} + S^{-2} \rightarrow \underline{MnS} \text{ (Whitish).}$$
$$Zn^{+2} + S^{-2} \rightarrow \underline{ZnS} \text{ (White).}$$

It will be noted that Fe^{+2} is considered to be present. This is because sulfide ions reduce Fe^{+3} to Fe^{+2} when Group II or Group III ions are present. In alkaline solutions ferric sulfide precipitates if Fe^{+3} is present:

$$2Fe^{+3} + 3S^{-2} \rightarrow \underline{Fe_2S_3} \text{ (Black).}$$

Step 25. Solution of the Precipitate for Group III. The precipitate consists of the hydroxides of Al^{+3}, Cr^{+3}, and Fe^{+3} plus the sulfides of Fe^{+2}, Co^{+2}, Ni^{+2}, Mn^{+2}, and Zn^{+2}. All of these solids are easily dissolved by 6 M HCl except CoS and NiS. These two sulfides are peculiar in that they do not precipitate in acid solution but, once precipitated, they do not readily dissolve completely without the aid of the oxidizing power of HNO_3. Therefore both HCl and HNO_3 are employed to bring the precipitate into complete solution. The precipitate will be in the bottom of a test tube.

Add 20 drops (1 ml) of 6 M HCl solution. Stir for about 30 seconds. Observe the solution closely. If there is any black precipitate undissolved, cobalt or nickel or both are almost certain to be present. If there is no black undissolved precipitate, proceed with Step 26. If there is some black undissolved precipitate, add 6 drops of concentrated HNO_3, transfer to a small beaker, and boil gently until only 4 or 5 drops of liquid remain. Transfer the solution to a test tube with a medicine dropper, add 5 drops of water to the beaker, rinse, and add this solution to the first. There will be a considerable amount of sulfur suspended in the solution. Centrifuge and transfer the clear solution to a small beaker. The solution may contain Al^{+3}, Cr^{+3}, Fe^{+3}, Mn^{+2}, Co^{+2}, Ni^{+2}, and Zn^{+2} and is ready for Step 26.

EQUATIONS FOR REACTIONS IN STEP 25. The addition of HCl dissolves the solids except for CoS and NiS. The reactions are between the H_3O^+, from HCl and water, and the solids:

$$Al(OH)_3 + 3H_3O^+ \rightarrow Al^{+3} + 6H_2O.$$
$$Cr(OH)_3 + 3H_3O^+ \rightarrow Cr^{+3} + 6H_2O.$$
$$Fe(OH)_3 + 3H_3O^+ \rightarrow Fe^{+3} + 6H_2O.$$
$$FeS + 2H_3O^+ \rightarrow Fe^{+2} + H_2S\uparrow + 2H_2O.$$
$$MnS + 2H_3O^+ \rightarrow Mn^{+2} + H_2S\uparrow + 2H_2O.$$
$$ZnS + 2H_3O^+ \rightarrow Zn^{+2} + H_2S\uparrow + 2H_2O.$$

Treatment with HNO_3 in addition to HCl dissolves sulfides of nickel and cobalt by oxidizing the few sulfide ions in the saturated solutions of CoS and NiS. The sulfide ions are formed when very small amounts of nickel and cobalt sulfides dissolve:

$$CoS \rightleftharpoons Co^{+2} + S^{-2}.$$
$$NiS \rightleftharpoons Ni^{+2} + S^{-2}.$$

Sulfide is then oxidized by the ions produced when nitric acid dissolves in water:

$$3S^{-2} + 2NO_3^- + 8H_3O^+ \rightarrow 2NO\uparrow + 12H_2O + \underline{3S} \text{ (White or yellow).}$$

As the sulfide is oxidized, the concentrations of nickel and cobalt ions are increased until the precipitates are completely dissolved.

The nitric acid treatment also oxidizes all Fe^{+2} to Fe^{+3}:

$$3Fe^{+2} + NO_3^- + 4H_3O^+ \rightarrow 3Fe^{+3} + NO\uparrow + 6H_2O.$$

Step 26. Separation of the Aluminum Division from **the Nickel Division.** The solution from Step 25 may contain the ions of the group shown as products in the equations for reactions in Step 25.

Add 20 drops (1 ml) of 6 M NaOH and 5 drops of 3% H_2O_2 solutions. Stir; then place the beaker on a screen and heat the screen half an inch or so away from the beaker, just enough to boil the solution gently. Keep the eyes well away, and wear eye shields! Strongly alkaline solutions tend to "bump" and spatter when boiled. "Bumping" can be reduced by constant stirring, and rubbing the bottom of the hottest part of the beaker. Take the flame away, add a drop of

3% H_2O_2 solution, boil, remove the flame, add another drop of 3% H_2O_2, and repeat until 5 drops have been added. The total amount of 3% H_2O_2 solution added now amounts to 10 drops. The boiling should not be continued more than 30 seconds between drops or after the last drop of H_2O_2 is added. Transfer the solution and precipitate to a test tube for centrifuging. Use 2 or 3 drops of distilled water to rinse out the beaker and complete the transfer. Centrifuge and decant the clear solution into a small test tube. Wash the precipitate twice, each time with 10 drops of distilled water, stirring the precipitate up well both times before centrifuging. Add the first wash solution to the test tube containing the main solution; discard the second wash solution. Make the solution acid with concentrated HNO_3, added a drop at a time, stirring after each drop is added, and testing with litmus after each drop is stirred in. This acid solution is ready for Step 35. If the strongly basic solution were not made acid at this time it would dissolve glass from the test tube and give a positive test for aluminum later. The precipitate may consist of $Fe(OH)_3$, $Co(OH)_3$, $Ni(OH)_2$, and MnO_2. It is in a test tube ready for Step 27.

EQUATIONS FOR REACTIONS IN STEP 26. The addition of NaOH increases the hydroxide ion concentration to a point where the solubility product constants are exceeded for the hydroxides of all of the ions in the group:

$$Fe^{+3} + 3OH^- \rightarrow \underline{Fe(OH)_3} \text{ (Red-brown).}$$
$$Co^{+3} + 3OH^- \rightarrow \underline{Co(OH)_3} \text{ (Black).}$$

If any cobalt remains unoxidized:

$$Co^{+2} + 2OH^- \rightarrow \underline{Co(OH)_2} \text{ (Pink).}$$
$$Ni^{+2} + 2OH^- \rightarrow \underline{Ni(OH)_2} \text{ (Light green).}$$
$$Mn^{+2} + 2OH^- \rightarrow \underline{Mn(OH)_2} \text{ (White to brown).}$$
$$Al^{+3} + 3OH^- \rightarrow \underline{Al(OH)_3} \text{ (White).}$$
$$Cr^{+3} + 3OH^- \rightarrow \underline{Cr(OH)_3} \text{ (Green).}$$
$$Zn^{+2} + 2OH^- \rightarrow \underline{Zn(OH)_2} \text{ (White).}$$

If further addition of hydroxide is made, the hydroxide ion concentration is increased until the amphoteric hydroxides dissolve, forming negative ions:

$$Al(OH)_3 + OH^- \rightarrow Al(OH)_4^-.$$
$$Cr(OH)_3 + OH^- \rightarrow CrO_2^- + 2H_2O.$$
$$Zn(OH)_2 + 2OH^- \rightarrow ZnO_2^{-2} + 2H_2O.$$

Peroxide added oxidizes Cr^{+3} to CrO_4^{-2}, Mn^{+2} to MnO_2, and Co^{+2} to Co^{+3}. Oxidation of the chromium is essential in order to get all of the chromium into solution because $Cr(OH)_3$ does not dissolve completely, as indicated by the equation above. In the higher oxidation state chromium is more like a nonmetal in forming the very soluble chromate ion. Oxidation of manganese is desirable because MnO_2 is much less soluble than $Mn(OH)_2$. Oxidation of cobalt is incidental, but it occurs. The equations for these reactions are:

$$2CrO_2^- + 2OH^- + 3H_2O_2 \rightarrow 2CrO_4^{-2} + 4H_2O.$$
$$Mn(OH)_2 + H_2O_2 \rightarrow 2H_2O + \underline{MnO_2} \text{ (Black or dark brown).}$$
$$2Co(OH)_2 + H_2O_2 \rightarrow \underline{2Co(OH)_3} \text{ (Black).}$$

Step 27. Solution of the Nickel Division Precipitate. The hydroxides of Co^{+3}, Fe^{+3}, and Ni^{+2} are dissolved completely and MnO_2 is partially dissolved with HNO_3. H_2O_2 is added to reduce MnO_2 to Mn^{+2}, which completes the solution of manganese.

Add 10 drops (0.5 ml) of concentrated HNO_3 to the precipitate from Step 26. Mix well; then add 3 drops, stirring after each drop is added, of 3% H_2O_2 solution. Place the test tube in boiling water for 1 minute. Inspect the contents of the tube. If all of the precipitate is not redissolved, add another drop or two of H_2O_2, mix, and warm again for 1 minute. The solution must be reduced in volume for the next step. Transfer it from the test tube to a small beaker, rinsing out the test tube with a few drops of concentrated HNO_3. Boil the solution in the beaker until about half the original volume remains. Do not boil dry. Boiling for a few minutes decomposes all H_2O_2. Now add 12 to 15 drops of concentrated HNO_3. The solution is now ready for Step 28.

EQUATIONS FOR REACTIONS IN STEP 27. Nitric acid dissolves the hydroxides:

$$Co(OH)_3 + 3H_3O^+ \rightarrow Co^{+3} + 6H_2O.$$
$$Fe(OH)_3 + 3H_3O^+ \rightarrow Fe^{+3} + 6H_2O.$$
$$Ni(OH)_2 + 2H_3O^+ \rightarrow Ni^{+2} + 4H_2O.$$

Peroxide ions from H_2O_2 then reduce Co^{+3} to Co^{+2} and MnO_2 to Mn^{+2}:

$$2Co^{+3} + O_2^{-2} \rightarrow 2Co^{+2} + O_2\uparrow.$$
$$MnO_2 + O_2^{-2} + 4H_3O^+ \rightarrow Mn^{+2} + O_2\uparrow + 6H_2O.$$

Hydrogen peroxide decomposes when boiled:

$$2H_2O_2 \rightarrow H_2O + O_2\uparrow.$$

Step 28. Separation of Manganese from Other Elements of the Nickel Division. The reaction employed is the formation of very insoluble MnO_2 in a strong nitric acid solution by the action of chlorate ion on Mn^{+2}.

Heat the solution from Step 27 to near boiling and add 0.15 g of $KClO_3$ in about 8 to 10 equal portions. While heating and adding $KClO_3$, stir the solution constantly. After about a minute of heating remove the beaker from the heating surface and inspect the solution. If a dark brown precipitate appears, it is probably MnO_2. If there is no precipitate at all, no manganese was in the sample being analyzed and Step 29 need not be done. If there is a precipitate, transfer solution and precipitate together to a test tube, centrifuge, and decant the clear solution and place it in a small beaker. It is then ready for Step 30. Wash the precipitate twice, each time with 5 drops of hot distilled water. Discard the washings. The precipitate is now ready for Step 29.

Equation for the Reaction in Step 28. The only significant reaction is the oxidation of Mn^{+2} to MnO_2 by chlorate. Chlorine dioxide is also formed, but it either boils away or reacts with water to yield a variety of products:

$$Mn^{+2} + 2ClO_3^- \rightarrow 2ClO_2 \uparrow + \underline{MnO_2} \text{ (Black or brown).}$$

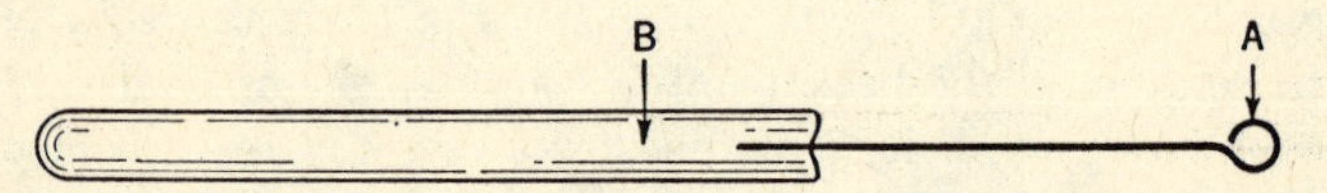

Fig. 5. Apparatus for a Bead Test. A piece of wire, preferably platinum, A, is sealed into a glass rod, B, and a small loop is made in the wire as shown.

Step 29. The Tests for Manganese. *Test 1.* A quick test for manganese, made on MnO_2, is called the *bead test.* Make a small loop on the end of a piece of platinum wire, as shown in Figure 5, heat it red-hot, touch it to powdered sodium carbonate, Na_2CO_3, heat again in the flame, and touch it to sodium carbonate again. Repeat until a bead of molten sodium carbonate is formed in the loop of wire. While the bead is hot, touch it to a bit of MnO_2 and heat the bead again in the flame for a minute or two. Remove it from the flame, cool it, and inspect it for the green color of MnO_4^{-2} (manganate).

A greenish bead confirms the presence of manganese, which was suspected if a precipitate resembling MnO_2 was formed in the first place.

Test 2. An equally sensitive test, perhaps more reliable, is the conversion of MnO_2 to Mn^{+2}, then oxidation of Mn^{+2} to purple MnO_4^- (permanganate) in solution, using $NaBiO_3$ (sodium bismuthate) as the oxidizing agent. To the MnO_2 in the test tube add 6 to 8 drops of 3 M HNO_3 solution and a drop of 3% H_2O_2 solution. Stir, heat in hot water, add another drop of H_2O_2, stir, and heat until the precipitate is dissolved. Transfer the solution to a small beaker, boil at least 30 seconds to destroy excess H_2O_2, and cool. All peroxide must be decomposed by boiling because it will interfere with the test for manganese by acting as a reducing agent on MnO_4^-, which is to be prepared. Transfer the solution to a test tube, add a small bit of solid $NaBiO_3$ (sodium bismuthate), stir, centrifuge, and inspect the solution for a purple color which will confirm the presence of manganese.

EQUATIONS FOR REACTIONS IN STEP 29. *Test 1.* The reactions occurring in the bead test are probably rather complicated. Oxygen of the air and sodium carbonate react at this high temperature to produce a green manganate, probably Na_2MnO_4, dissolved in sodium carbonate. A possible equation is:

$$2MnO_2 + O_2 + 2CO_3^{-2} \rightarrow 2MnO_4^{-2} + 2CO_2\uparrow.$$

Test 2. In the second test the first reaction is the dissolving of MnO_2. The equation is the same as is given in Step 27:

$$MnO_2 + O_2^{-2} + 4H_3O^+ \rightarrow Mn^{+2} + O_2\uparrow + 6H_2O.$$

The second reaction is the oxidation of Mn^{+2} by bismuthate:

$$2Mn^{+2} + 5BiO_3^- + 14H_3O^+ \rightarrow 2MnO_4^- + 5Bi^{+3} + 21H_2O.$$

Step 30. The Separation of Iron from Cobalt and Nickel. In a high concentration of NH_3, Ni^{+2} and Co^{+2} form complex ions which are soluble, while Fe^{+3} forms insoluble $Fe(OH)_3$.

The solution from Step 28 is strongly acid with HNO_3. Boil it down until it is no more than half its original volume or until a solid begins to form. The solid is KCl from decomposition of $KClO_3$. Add distilled water to double the volume, cool, and, while stirring, add concentrated ammonia solution drop by drop. A great deal of heat will be generated by neutralization of the acid with NH_3. When the solution first tests alkaline with litmus, add 8 drops more ammonia solution. Mix thoroughly, transfer the mixture to a small test tube, centrifuge, and decant the solution into a small beaker. It is now ready for Step 32. The precipitate, if any, is reddish or brownish $Fe(OH)_3$. Wash it with 10 drops of distilled water and discard the washing. The precipitate is ready for Step 31.

EQUATIONS FOR REACTIONS IN STEP 30. The first reaction is precipitation of the hydroxides of the metals, brought about by increasing the hydroxide ion concentration with ammonia to a point where the solubility product constants for the hydroxides are exceeded:

$$Fe^{+3} + 3OH^- \rightarrow \underline{Fe(OH)_3} \text{ (Red-brown).}$$
$$Ni^{+2} + 2OH^- \rightarrow \underline{Ni(OH)_2} \text{ (Greenish).}$$
$$Co^{+2} + 2OH^- \rightarrow \underline{Co(OH)_2} \text{ (Pink).}$$

The latter two hydroxides dissolve in an excess of ammonia while ferric hydroxide remains insoluble:

$$Ni(OH)_2 + 6NH_3 \rightarrow Ni(NH_3)_6^{+2} + 2OH^-.$$
$$Co(OH)_2 + 6NH_3 \rightarrow Co(NH_3)_6^{+2} + 2OH^-.$$

Step 31. The Tests for Iron. The precipitated $Fe(OH)_3$ is dissolved in HCl solution. Then the Fe^{+3} can be tested for by two different methods. One is to add thiocyanate ion (SCN^-), which produces an intense red-colored solution. The other is to add hexacyanoferrate(II) (ferrocyanide ion), in which case a dark-blue precipitate, $KFe[Fe(CN)_6]$, forms if iron is present. The first of these two tests, described below, is usually sufficient.

To the precipitate of $Fe(OH)_3$ from Step 30 add 1 ml (20 drops) of 3 M HCl solution. If the precipitate does not dissolve at once, heat it in hot water until it does. Cool and add 4 drops of 0.5 M potassium thiocyanate solution. If a deep-red solution is produced, the presence of iron is confirmed, but a pink color is not significant, since there is usually a trace of iron in the reagents used, sufficient to produce a pink color.

EQUATIONS FOR REACTIONS IN STEP 31. The first reaction is the dissolving of $Fe(OH)_3$ with acid:

$$Fe(OH)_3 + 3H_3O^+ \rightarrow Fe^{+3} + 6H_2O.$$

The reaction for the final test for iron is:

$$Fe^{+3} + SCN^- \rightarrow FeSCN^{+2} \text{ (Red solution).}$$

There is no evidence that $Fe(SCN)_6^{-3}$ is ever formed in water solutions. Early work on the structure of the red compound indicated that $Fe(SCN)_6^{-3}$ and other compounds such as $Fe(SCN)_4^-$ might exist. More recent work shows that water molecules compete quite strongly with SCN^-, preventing the latter from occupying more than one position on each ferric ion in any but very high concentrations of SCN^-.

Step 32. Preparation of the Solution for the Cobalt and Nickel Tests. The strong ammonia solution containing cobalt and nickel must be boiled to reduce the concentration of ammonia. Then a slightly acid buffering solution (see p. 36), acetic acid plus potassium acetate, is added to prepare for the final tests for cobalt and nickel.

Boil down the solution from Step 30 to half of the original volume. Add 5 drops of 2 M $KC_2H_3O_2$ (potassium acetate) and 6 M acetic acid until the solution tests acid to litmus. Then add 3 drops more of the 6 M acetic acid. Divide the solution, two-thirds for the cobalt test and one-third for the nickel test, placing each portion in a separate test tube. The solutions are now ready for Steps 33 and 34.

EQUATIONS FOR REACTIONS IN STEP 32. The main reaction is the removal of NH_3 from the nickel and cobalt complexes by reaction with acetic acid:

$$Co(NH_3)_6^{+2} + 6HC_2H_3O_2 \rightarrow Co^{+2} + 6NH_4^+ + 6C_2H_3O_2^-.$$
$$Ni(NH_3)_6^{+2} + 6HC_2H_3O_2 \rightarrow Ni^{+2} + 6NH_4^+ + 6C_2H_3O_2^-.$$

More acetate ions are added as potassium acetate, $KC_2H_3O_2$, to assure sufficient buffering against hydronium ions. Acetic acid is added to make the solution faintly acid in the high acetate concentration.

Step 33. The Test for Nickel. Dimethylglyoxime is added to one-third of the solution from Step 32. If nickel is present a bright-red precipitate forms which is a complex compound in which the nickel atom is bonded to the dimethylglyoxime molecule by both ionic and coordinate covalent bonds.

One-third of the solution from Step 32 should be about 5 drops. To this volume add 1 drop of concentrated ammonia solution and 15 drops of water. Stir thoroughly and add 4 drops of dimethylglyoxime solution. Stir again and set the test tube aside for 15 minutes or so. Inspect the test tube to see if a red precipitate has formed, confirming the presence of nickel.

EQUATION FOR THE REACTION IN STEP 33. A little ammonia is added to reduce the acidity slightly. Then dimethylglyoxime reacts with nickel ions to produce the red precipitate. No ion other than Ni^{+2} will give the same-colored precipitate with this reagent under these conditions.

$$2H_2O + Ni^{+2} + 2CH_3C(NOH)C(NOH)CH_3 \rightarrow$$
$$2H_3O^+ + \underline{Ni[CH_3C(NO)C(NOH)CH_3]_2} \text{ (Red)}.$$

Step 34. The Tests for Cobalt. Several reagents are known which will identify cobalt in the presence of nickel.

Test 1. A bead test performed as described in the quick test for manganese, Step 29, is very sensitive. Instead of sodium carbonate, Na_2CO_3, borax, $Na_2B_4O_7 \cdot 10H_2O$, is melted into the test bead.

Make a wire loop as described in Step 29. Heat the loop in a flame until it is red-hot, then touch the hot loop to powdered borax. Some borax will stick to the wire. Melt the borax in the flame, touch the loop to powdered borax again, melt this in the flame, and continue adding to the borax on the loop until the loop is filled with molten borax to form a bead. While the bead is hot, touch it to any solid obtained as a precipitate in the nickel division, or to the original sample, if it is solid. A small amount of material will adhere to the bead. Now heat the bead in a flame for a minute or so, cool it, and inspect it. If the bead is a deep blue, the presence of cobalt is confirmed. A very dark-blue or black bead is confusing. Remove such a bead from the wire loop, break it up into small pieces, and test some of the small pieces with a fresh borax bead. A bead more dilute in cobalt is thus obtained, which, if blue, indicates the presence of cobalt. This test is so sensitive that some nickel salts, supposedly pure, have sufficient cobalt in them to give some blue color to a borax bead.

Test 2. Add 6 M H_2SO_4 to the solution from Step 32 until the solution is acid. Add 0.1 g or so of sodium fluoride, NaF, and 0.5 ml of some organic solvent such as amyl alcohol. Then saturate the solution with a gram or so of solid ammonium thiocyanate, NH_4SCN. Shake the mixture vigorously at least 20 seconds; then observe the organic solvent layer. Cobalt, if present, forms a blue or blue-green complex with the probable composition $Co(SCN)_4^{-2}$, which dissolves in the solvent layer.

Test 3. This test is simple and reliable. To the solution from Step 32 add a volume of 6 M KNO_2 (potassium nitrite) solution equal to that of the test solution at this point. Warm the solution in boiling water until it is near boiling and set it aside to cool. A yellow precipitate, which forms slowly with a small amount of cobalt, confirms the presence of cobalt. The precipitate is $K_3Co(NO_2)_6$, tripotassium hexanitritocobaltate(III).

EQUATIONS FOR REACTIONS IN STEP 34. The reactions occurring in Test 1, the bead test, are somewhat obscure. It is probable that some sort of blue cobalt borate is formed. Cobalt oxides are added to glass and ceramics in small amounts to produce a rich blue color. A possible equation is:

$$CoO + Na_2B_4O_7 \rightarrow Co(BO_2)_2 + 2NaBO_2.$$

In Test 2 the reactions are better understood. Fluoride is added to form a stable, colorless complex with Fe^{+3} and prevent traces of iron from producing red $Fe(SCN)^{+2}$, which would interfere with the cobalt test:

$$Fe^{+3} + 6F^- \rightarrow FeF_6^{-3} \text{ (Colorless).}$$

Then thiocyanate (SCN^-) ions react with Co^{+2} ions to produce the characteristic blue complex which is soluble in organic solvents:

$$Co^{+2} + 4SCN^- \rightarrow Co(SCN)_4^{-2} \text{ (Blue).}$$

In Test 3 the reaction is one involving oxidation of Co^{+2} to Co^{+3} with subsequent precipitation of an insoluble coordination compound. Hydronium ions furnished by acetic acid are needed for the oxidation reaction:

$$Co^{+2} + NO_2^- + 2H_3O^+ \rightarrow NO\uparrow + Co^{+3} + 3H_2O.$$

$$Co^{+3} + 3K^+ + 6NO_2^- \rightarrow \underline{K_3Co(NO_2)_6} \text{ (Yellow).}$$

Step 35. Separation of Aluminum from Chromium and Zinc. Whether or not aluminum is separated earlier in the Group III analysis, it is always separated as the hydroxide from a solution buffered with ammonium ions, NH_4^+, and made slightly alkaline with NH_3 solution. The solution from Step 26 was made acid with HNO_3 and is in a small beaker.

Add concentrated ammonia solution slowly, stirring steadily and testing with litmus by removing the stirring rod and touching it to litmus paper occasionally. The heat of reaction between nitric acid and ammonia will cause rapid boiling which may blow liquid out of the beaker unless the ammonia solution is added very slowly and with constant stirring. When the solution tests alkaline, if aluminum is present, a whitish, almost invisible precipitate will form. Centrifuge whether a precipitate can be seen or not. Inspect the centrifuge tube for the presence of a white precipitate; if there is one, aluminum is probably present. If the precipitate is greenish, it may in part be $Cr(OH)_3$, all of which should have been converted to chromate in Step 26. Whatever the color of the precipitate, decant the solution into a test tube and save it for Step 37. Wash the solid 3 times with 5 drops of hot distilled water, discarding the washings. The precipitate is now ready for Step 36.

EQUATIONS FOR REACTIONS IN STEP 35. The reaction of an acid such as HNO_3 with ammonia can be indicated by the equation:

$$H_3O^+ + NH_3 \rightarrow NH_4^+ + H_2O.$$

Ammonium ions prevent too high a concentration of hydroxide ions from developing in a solution when ammonia is added. As ammonia is added, the hydroxide ion concentration is increased, and when the solubility product constant for aluminum hydroxide is exceeded, a precipitate forms:

$$Al^{+3} + 3OH^- \rightarrow \underline{Al(OH)_3} \text{ (White)}.$$

If no ammonium ions were present to buffer the solution, the precipitate formed might dissolve in the high hydroxide ion concentration, forming $Al(OH)_4^-$, which is very soluble.

While aluminum is precipitating, chromate ions, if present, remain unchanged and zinc ions form a soluble ammonia complex. Zinc hydroxide usually forms first as a white precipitate that somewhat resembles aluminum hydroxide:

$$Zn^{+2} + 2OH^- \rightarrow \underline{Zn(OH)_2} \text{ (White)}.$$

This material dissolves in an excess of ammonia:

$$Zn(OH)_2 + 4NH_3 \rightarrow Zn(NH_3)_4^{+2} + 2OH^-.$$

Cr^{+3} ions appear in the solution from Step 26 if there is any Cr^{+3} not oxidized or if there is undecomposed peroxide remaining in solution when nitric acid is added. In an alkaline solution peroxide oxidizes chromium ions to chromate, but in an acid solution peroxide reacts with dichromate, reducing it to Cr^{+3}. The dichromate ($Cr_2O_7^{-2}$) is produced when acid is added:

$$2CrO_4^{-2} + 2H_3O^+ \rightarrow Cr_2O_7^{-2} + 3H_2O.$$

Peroxide then reduces dichromate:

$$Cr_2O_7^{-2} + 3O_2^{-2} + 14H_3O^+ \rightarrow 2Cr^{+3} + 3O_2\uparrow + 21H_2O.$$

Step 36. The Test for Aluminum. This test depends on the fact that $Al(OH)_3$ adsorbs an organic dye, aluminon, to produce a bright-red lake. (A *lake* is any metal oxide or other insoluble compound on which a dye is adsorbed.) $Cr(OH)_3$ and $Zn(OH)_2$ do not adsorb aluminon. The green of $Cr(OH)_3$ may mask the red of the aluminum lake. Therefore, if the precipitate from Step 35 is green, it must be treated again as in Steps 25, 26, and 35.

To the precipitate from Step 35 add 5 drops of 3 M HNO_3 solution. All the precipitate should dissolve. If it does not, centrifuge and discard the solid which did not dissolve. To the clear solution add 2 or 3 drops of aluminon solution and add 6 M ammonia solution a drop at a time, stirring and testing with litmus until the solution is just barely alkaline. A red, flocculent (cloudlike) precipitate is a positive test for aluminum. It is easier to see a small precipitate if the suspension is centrifuged so the solid is concentrated at the bottom of the test tube.

EQUATIONS FOR REACTIONS IN STEP 36. There are only two chemical reactions in Step 36, that of the dissolving of $Al(OH)_3$ and that of reprecipitating it in the presence of the dye. The two equations are:

$$Al(OH)_3 + 3H_3O^+ \rightarrow Al^{+3} + 6H_2O.$$

$$Al^{+3} + 3OH^- \rightarrow \underline{Al(OH)_3} \text{ (White, colored red by the dye).}$$

No equation for the adsorption of the dye can be written. It is a physical rather than a chemical change, although a somewhat chemical type of formation of bonds may take place in the adsorption process. Aluminon is the ammonium salt of aurintricarboxylic acid. Its formula is:

$$(C_6H_3OHCOONH_4)_2CC_6H_3OHCOONH_4.$$

Step 37. Separation of Chromium from Zinc. If the solution from Step 35 is not yellow and if the hydroxide precipitate from Step 35 is not green, no chromium can be present and Steps 37 and 38 can be omitted from the procedure. The yellow color of chromate is very sensitive, but not completely conclusive. A confirming test should be made.

To the yellow solution from Step 35 add 6 M acetic acid until the solution is barely acid. Then add 5 drops of 0.2 M $BaCl_2$ (barium chloride) solution. Stir the mixture and centrifuge. Decant the solution into a test tube for Step 39. Wash the precipitate with two 5-drop portions of distilled water. Discard the water. The precipitate is now ready for Step 38. It may be all $BaCrO_4$, or $BaCrO_4$ mixed with $BaSO_4$. Sulfate may be present as a result of air oxidation of sulfide earlier in the procedure.

EQUATIONS FOR REACTIONS IN STEP 37. The first reaction, making the ammonia solution acid with acetic acid, is:

$$NH_3 + HC_2H_3O_2 \rightarrow NH_4^+ + C_2H_3O_2^-.$$

Then the tetramminezinc ions are converted to zinc ions:

$$Zn(NH_3)_4^{+2} + 4HC_2H_3O_2 \rightarrow Zn^{+2} + 4NH_4^+ + 4C_2H_3O_2^-.$$

On making the solution acid, chromate may be partially converted to dichromate by hydronium ions in the acetic acid solution:

$$2CrO_4^{-2} + 2H_3O^+ \rightleftharpoons Cr_2O_7^{-2} + 3H_2O.$$

Finally, barium ions, from barium chloride, form insoluble $BaCrO_4$, with chromate ions not completely converted to dichromate:

$$Ba^{+2} + CrO_4^{-2} \rightarrow \underline{BaCrO_4} \text{ (Yellow)}.$$

Step 38. The Test for Chromium. In this step a chromate is converted to dichromate and then, with hydrogen peroxide, H_2O_2, to a deep-blue-colored substance, CrO_5, soluble in ether and water. It is not very stable in either solvent, but is much more stable in ether than in water. Since ether is highly flammable, make sure that no flame is burning nearby when this test is made.

To the solid from Step 37 add 3 drops of 3 M HNO_3, warm, and stir. Do not heat to boiling. Cool with cold running water to as low a temperature as the tap water. Add about 12 drops of cooled distilled water, stir, and add enough ether to make a layer about $\frac{1}{4}$ inch thick above the water. Add 1 drop of 3% H_2O_2 solution and shake immediately, with the thumb over the top of the tube. Hold the tube upright when removing the thumb from the top of the tube after shaking. The formation of a deep-blue color in the ether layer is confirmation of the presence of chromium.

EQUATIONS FOR REACTIONS IN STEP 38. The first reaction is the dissolving of $BaCrO_4$ with nitric acid solution which simultaneously converts chromate to dichromate:

$$2BaCrO_4 + 2H_3O^+ \rightarrow 2Ba^{+2} + Cr_2O_7^{-2} + 3H_2O.$$

The final test for chromium is the conversion by peroxide of dichromate to CrO_5, chromium peroxide, which is not stable for more than a few minutes.

$$Cr_2O_7^{-2} + 4O_2^{-2} + 10H_3O^+ \rightarrow 2CrO_5 + 15H_2O.$$

CrO_5 is more stable when cold than at high temperatures. It is also less stable in high concentration of nitric acid than in dilute acid, but some acid is necessary for its formation. CrO_5 is extracted from the water layer by ether in order to prevent contact with HNO_3, which is not soluble in ether. The CrO_5 soon decomposes into Cr^{+3} and other products. Two possible equations are:

$$4CrO_5 + 12H_3O^+ \rightarrow 2Cr^{+3} + 7O_2\uparrow + 18H_2O.$$
$$2CrO_5 \rightarrow 2Cr^{+3} + 2O_2\uparrow + 3O_2^{-2}.$$

Step 39. The Test for Zinc. There are at least two excellent tests for zinc that can be made on the solution from Step 35 or Step 37. One test is made by adjusting the pH and adding a source of sulfide ions such as ammonium sulfide solution. A white precipitate (ZnS) is proof of the presence of zinc. It often happens in making this test that some sulfide is oxidized by nitrate to sulfur, which is white when freshly precipitated. To be sure that the white precipitate is ZnS and not sulfur, it must be dissolved in acid and reprecipitated. Sulfur is not soluble in acid, whereas ZnS is. This reprecipitation process is rather lengthy, but it is necessary if the operator is to be certain of the test.

Ferrocyanide, $Fe(CN)_6^{-4}$, better called by its proper name, hexacyanatoferrate(II), when added to a slightly acid solution containing zinc ions produces a whitish precipitate which is quite satisfactory as a test for zinc.

To the solution from either Step 35 or Step 37, add 6 M HCl dropwise, testing with litmus until the solution is barely acid. Then add 5 drops of 0.2 M $K_4Fe(CN)_6$ (tetrapotassium hexacyanatoferrate(II), and stir. If a white or off-white precipitate forms, the presence of zinc is confirmed.

Equations for Reactions in Step 39. The solution from Step 35 may contain $Zn(NH_3)_4^{+2}$. Adding HCl solution provides H_3O^+ (hydronium ions) to convert the ammonia complex to zinc ions:

$$Zn(NH_3)_4^{+2} + 4H_3O^+ \rightarrow Zn^{+2} + 4NH_4^+ + 4H_2O.$$

The solution from Step 37 contains Zn^{+2} already. The final test for zinc is the reaction of Zn^{+2} with $Fe(CN)_6^{-4}$ and K^+:

$$3Zn^{+2} + 2K^+ + 2Fe(CN)_6^{-4} \rightarrow \underline{K_2Zn_3[Fe(CN)_6]_2} \text{ (White)}.$$

Review Questions and Problems

1. Write out Chart 4 from memory.
2. From Chart 4 write equations for all reactions in all the steps of the analysis of Group III.
3. For the following sets of ions, each set contained in a single solution, outline the steps needed to separate and confirm the presence of each ion:

   a. Ag^+, Cd^{+2}, Al^{+3}, and Co^{+2}
   b. Cr^{+3}, Fe^{+3}, Mn^{+2}, and Zn^{+2}
   c. Cr^{+3} and Ni^{+2}
   d. Hg_2^{+2}, Sb^{+3}, and Mn^{+2}
   e. Pb^{+2} and Zn^{+2}
   f. Mn^{+2} and Co^{+2}

4. a. Why is fluoride ion added when cobalt is tested for with thiocyanate?
   b. Why do zinc ions yield, with ammonia, a white precipitate which dissolves when more ammonia solution is added?
   c. Why is nitric acid added in making the final test for chromium?
   d. Why is NH_4Cl added when aluminum is separated from zinc?
   e. Why does MnO_2 dissolve in nitric acid when H_2O_2 is present, but not in nitric acid alone?
   f. Why may there be chromium(III) in the solution from Step 26?
5. To a colorless solution containing only elements in Group III, NaOH solution is added slowly. A precipitate forms, but on further addition of NaOH it dissolves.
   a. What elements are absent?
   b. What elements may be present?
6. Choose at random any pair of ions in Group I, II, or III and write the formula or formulas of reagents necessary for separating the two.
7. Do the same thing as in problem 6, except that instead of two ions, choose three or four or more ions.
8. A solid mixture containing ions of Group III dissolves in water, producing a darkish solution. When NH_4Cl and NH_3 solution are added a dark green precipitate is formed, but when $(NH_4)_2S$ is added still more precipitate forms, which makes the whole suspension appear black. The precipitate is treated with acid, and sodium hydroxide and peroxide are added. The solution becomes yellow, but a dark solid remains undissolved.
   a. At this point what ion or ions are indicated as probably present?

   The solid is dissolved in acid and the resulting solution is treated with $KClO_3$; no precipitate forms. Ammonia is added and no precipitate forms; acetic acid, acetate, and KNO_2 are added without any precipitate forming.

   b. What ions are absent?
   c. What evidence indicates that a positive test for nickel is to be expected?

6

Analysis of Group IV, the Calcium Group

As with Group III, the analysis of Group IV can be accomplished by several different procedures. The usual reagent for precipitating Group IV ions as carbonates is $(NH_4)_2CO_3$ plus a buffer solution of NH_3 and NH_4Cl. Ammonium ions limit the hydroxide ion concentration. Thus the hydronium ion concentration is maintained great enough so that the carbonate ion concentration never becomes large enough to exceed the solubility product constant for $MgCO_3$. Magnesium ions therefore precipitate, not in Group IV, but in Group V along with Na^+, K^+, and NH_4^+, the last being present because it is added in the various previous procedures.

Step 40. Removal of Excess Ammonium Ions. Although ammonium ions are necessary for controlling the carbonate ion concentration, previous steps have caused an accumulation of NH_4^+. In order to be able to control the NH_4^+ concentration accurately, it is therefore desirable, unless the sample is one containing only Group IV ions, to remove all ammonium ions and then add the amount needed to produce the concentration required. If the sample to be analyzed contains only ions of Groups IV and V, Step 40 can be omitted.

Place the solution from Step 24 in a casserole or a crucible. Add 15 drops of concentrated HNO_3. Place the casserole or crucible over a wire screen under a hood and boil the solution gently until all liquid is evaporated. Slowly raise the temperature until white smoke begins to form, and continue heating until the smoking ceases, but do not heat to red heat. The white smoke is ammonium salts being sublimed. Cool the vessel in air until it is below the boiling point of water. Add 3 drops of 6 M HCl solution and 10 drops of distilled water. Transfer to a test tube, rinsing the vessel twice with 10-drop portions of distilled water. Add the rinsings to the test tube.

CHART 5

Analysis of Group IV, the Calcium Group

The solution from Step 24 or a solution known to contain only elements of Group IV may contain any or all of the following:

Ba^{+2}, Sr^{+2}, Ca^{+2}, and Mg^{+2}.

Step 40 Add { HNO_3.
Boil dry.
Bake.
↓
Ba^{+2}, Sr^{+2}, Ca^{+2}, and Mg^{+2}

Step 41 Add { NH_4Cl, NH_3, $(NH_4)_2CO_3$.
Centrifuge.

Solid ↓ $\underline{BaCO_3}$, $\underline{SrCO_3}$, $\underline{CaCO_3}$

Solution ↓ Mg^{+2}, K^+, Na^+ (and NH_4^+ from Step 41) (for Step 49)

Step 42 Add { $HC_2H_3O_2$.
↓
Ba^{+2}, Sr^{+2}, Ca^{+2}

Step 43 Add { $NH_4C_2H_3O_2$, K_2CrO_4.
Centrifuge.

Solid ↓ $\underline{BaCrO_4}$

Step 44 Add { HCl.
Flame test, green

Solution ↓ Sr^{+2}, Ca^{+2}

Step 45 Add { NH_3, $(NH_4)_2CO_3$.
Centrifuge.

Solid ↓ $\underline{SrCO_3}$, $\underline{CaCO_3}$

Solution ↓ Discard.

Step 46 Add { HNO_3.
Centrifuge.

Solid ↓ $\underline{Sr(NO_3)_2}$

Solution ↓ Ca^{+2}

Step 47 Add $\begin{cases} H_2O, \\ (NH_4)_2SO_4. \end{cases}$
↓
$\underline{SrSO_4}$
White

Step 48 Add $\begin{cases} NH_3, \\ (NH_4)_2C_2O_4. \end{cases}$
↓
$\underline{CaC_2O_4}$
White
For flame test, add {HCl.
Reddish-orange

Remove any solids which might cause cloudiness by centrifuging and discarding the solid. Decant the solution into a test tube for Step 41.

EQUATION FOR THE REACTION IN STEP 40. Ammonium ions are partially removed by oxidation with nitric acid at boiling temperature. The equation for the reaction is:

$$NH_4^+ + NO_3^- \rightarrow N_2O \uparrow + 2H_2O.$$

Baking, after dryness, sublimes (evaporates) remaining ammonium salts and evaporates HNO_3, if any is present.

Step 41. Precipitation of Group IV Ions. The solution from Step 40 or a solution known to contain only elements of Groups IV and V may contain Ba^{+2}, Sr^{+2}, Ca^{+2}, Mg^{+2}, Na^+, and K^+ and will be in a test tube.

Add distilled water if needed to make up to 1 ml. Add 10 drops of 4 M NH_4Cl solution, make alkaline to litmus with 15 M NH_3 solution, and add 10 drops of 2 M $(NH_4)_2CO_3$ (ammonium carbonate) solution. Stir and heat in hot water at 60°–70° C. After about 5 minutes, centrifuge, and decant the solution into a test tube. This solution is ready for Step 49. Wash the precipitate with hot distilled water twice, using 10 drops of water each time. Discard the washings. The precipitate is ready for Step 42.

EQUATIONS FOR REACTIONS IN STEP 41. The main reactions are combination of CO_3^{-2} (carbonate ions) with the ions of Group IV:

$$Ba^{+2} + CO_3^{-2} \rightarrow \underline{BaCO_3} \text{ (White)}.$$
$$Sr^{+2} + CO_3^{-2} \rightarrow \underline{SrCO_3} \text{ (White)}.$$
$$Ca^{+2} + CO_3^{-2} \rightarrow \underline{CaCO_3} \text{ (White)}.$$

A small amount of $MgCO_3$ may precipitate at this point if the concentration of Mg^{+2} is relatively high. It will not interfere with subsequent tests in Group IV and will not be missed in the test for Mg^{+2} in Group V.

Step 42. Solution of the Precipitate of Group IV Carbonates. Before the ions of Group IV can be separated from each other, they must be brought into solution. Acetic acid ($HC_2H_3O_2$) is sufficiently strong to dissolve the carbonates from Step 41.

Add 5 drops of 6 M $HC_2H_3O_2$ (acetic acid) and stir. The precipitate should dissolve completely. More acid and distilled water may be added.

EQUATION FOR THE REACTION IN STEP 42. The main reaction — in fact, the only one — is that between the carbonate ions of the possible precipitates and hydronium (H_3O^+) from acetic acid solution in water:

$$CO_3^{-2} + 2H_3O^+ \rightarrow CO_2 \uparrow + 3H_2O.$$

This reaction reduces the carbonate ion concentration below that needed to exceed the solubility product constant for the carbonates of Group IV, and they therefore dissolve.

Step 43. Separation of Barium from Other Ions of Group IV. In a solution containing ammonium acetate, and acetic acid from Step 42, the chromate of barium is insoluble but the chromates of strontium and calcium remain soluble.

Add 3 drops of 3 M $NH_4C_2H_3O_2$ (ammonium acetate) solution and 4 drops of 0.5 M K_2CrO_4 (potassium chromate) solution. Stir and warm in boiling water 1 full minute. Cool the solution and precipitate, if any, and centrifuge. A yellow precipitate will form if Ba^{+2} was present in the sample. Transfer the solution to a test tube. It is ready for Step 45. The precipitate is ready for Step 44.

EQUATION FOR THE REACTION IN STEP 43. The only reaction is the combination of barium and chromate ions to form insoluble barium chromate:

$$Ba^{+2} + CrO_4^{-2} \rightarrow \underline{BaCrO_4} \text{ (Yellow)}.$$

Step 44. Detection of Barium. After some experience, the following test may be made successfully on the solution from Step 24 and in most cases on the original unknown solid or solution. It depends upon the fact that barium ions, when heated in a flame until they are in the vapor state, radiate energy in the form of light of a brilliant green hue. Only a few other elements do this. Thallium produces an intense-green flame, and copper salts produce a blue-green flame. With experience it is possible to learn to tell the difference between each of these three elements with flame tests.

To the precipitate from Step 43 add 5 drops of concentrated HCl solution and warm gently until the solution becomes green in color. Dip a clean platinum wire into the solution, hold it in a flame, and observe the color of the flame above the wire. If a smooth green flame color appears, the presence of barium is confirmed. A yellow precipitate may have been $SrCrO_4$ if a large amount of Sr^{+2} was present in the sample. In that case, barium being absent, the flame test would show only a red color. The green flame color of barium is much more persistent than the red of strontium, so that the green will show up after the red of strontium has faded, in case both strontium and barium are precipitated together.

EQUATION FOR THE REACTION IN STEP 44. The important chemical change in this step is the reduction of chromate ions by chloride ions in an acid solution. At the same time barium ions are liberated from solid barium chromate, leaving them free to evaporate much more readily as the chloride in the hot flame.

$$2BaCrO_4 + 16H_3O^+ + 6Cl^- \rightarrow 2Cr^{+3} + 3Cl_2\uparrow + 2Ba^{+2} + 24H_2O.$$

Step 45. Reprecipitation of Carbonates of Strontium and Calcium. The solution from Step 43 contains chromate ions, which must be eliminated so they will not interfere with future reactions. The carbonates of strontium and calcium are therefore precipitated and washed free of chromate before further treatment.

Neutralize the acetic acid in the solution from Step 43 by adding concentrated ammonia solution drop by drop, testing frequently with litmus. Add 10 drops of 2 M $(NH_4)_2CO_3$ (ammonium carbonate) solution. Stir well. If no precipitate forms at this point, no calcium or strontium is present and Steps 46, 47, and 48 can be omitted. If a precipitate forms, centrifuge, discard the solution, and wash the precipitate of $SrCO_3$ or $CaCO_3$ or both with 10 drops of hot distilled water as many times as necessary to obtain a completely colorless wash solution. Discard the wash solutions. The precipitate is in the bottom of the test tube ready for Step 46.

EQUATIONS FOR REACTIONS IN STEP 45. The reactions in this step are the same as those in Step 41, the combination of carbonate ions with strontium and calcium ions to form insoluble carbonates:

$$Sr^{+2} + CO_3^{-2} \rightarrow \underline{SrCO_3} \text{ (White).}$$
$$Ca^{+2} + CO_3^{-2} \rightarrow \underline{CaCO_3} \text{ (White).}$$

Step 46. Separation of Strontium from Calcium. This separation is based on the slight solubility of strontium nitrate, $Sr(NO_3)_2$, as compared with the great solubility of calcium nitrate, $Ca(NO_3)_2$.

Add concentrated HNO_3 solution a drop at a time, with stirring, to the precipitate from Step 45 until 30 or more drops have been added. Cool the test tube containing the solution for 5 minutes or more in cold water. Ice water is preferable. Centrifuge and decant the solution, which may contain calcium nitrate, into a small beaker. The solution is ready for Step 48. The precipitate, if any, is ready for Step 47.

EQUATIONS FOR REACTIONS IN STEP 46. The first reaction is that of hydronium ions (H_3O^+) from nitric acid in solution with carbonate ions in the precipitates of calcium and strontium carbonates:

$$SrCO_3 + 2H_3O^+ \rightarrow Sr^{+2} + CO_2 + 3H_2O.$$
$$CaCO_3 + 2H_3O^+ \rightarrow Ca^{+2} + CO_2 + 3H_2O.$$

The second reaction in this step is the combination of strontium and nitrate ions to form the precipitate $Sr(NO_3)_2$:

$$Sr^{+2} + 2NO_3^- \rightarrow \underline{Sr(NO_3)_2} \text{ (White)}.$$

Step 47. The Test for Strontium. If there is no precipitate in Step 46, strontium is absent, but if there is a precipitate a further test is advisable. A very reliable test is the flame test, similar to that made for barium in Step 44. Like the flame test for barium, this test may be made on the solution from Step 24 or on the original sample, after the operator becomes familiar with the characteristics of the flame tests for all the common elements that produce colored flames.

Add 10 to 15 drops of distilled water to the precipitate from Step 46. Heat in a flame until near boiling. Dip a well-cleaned platinum wire into the solution and bring it to the edge of a flame from a bunsen burner. A smooth red flame above the wire, which fades quickly, is assurance of the presence of strontium in the sample.

A further test for strontium can be made by precipitating strontium sulfate, $SrSO_4$, from the hot solution used for the flame test above. To do this, add 1 drop of 0.2 M $(NH_4)_2SO_4$ (ammonium sulfate) solution to this solution. The formation of a white precipitate, $SrSO_4$, is an indication that strontium was present in the unknown sample.

EQUATION FOR THE REACTION IN STEP 47. Ordinarily the flame test for strontium is a sufficiently reliable test and there are no reactions in flame tests that can be written as equations. The equation for the precipitation of $SrSO_4$ is:

$$Sr^{+2} + SO_4^{-2} \rightarrow \underline{SrSO_4} \text{ (White)}.$$

Step 48. The Test for Calcium. This test depends on the fact that calcium oxalate, CaC_2O_4, is a very insoluble white solid that precipitates out of a solution that is slightly basic with ammonia.

The solution from Step 46 is in a small beaker. It may contain calcium ions and it will be strongly acid with nitric acid. Add concentrated ammonia solution slowly down the side of the beaker, stirring fairly rapidly. Keep the beaker turned so that if the heat of reaction causes spattering no one will be injured by flying drops of nitric acid. Wear eye shields! When the acid is completely neutralized as indicated by a drop of solution on litmus paper from the stirring rod, add 1 more drop of ammonia solution. Then add 15 drops of 0.3 M ammonium oxalate solution, stir for about 30 seconds, and look closely for the appearance of a white, crystalline precipitate. The precipitate is calcium oxalate, CaC_2O_4, which may take up to 20 minutes to appear if only a small amount of calcium is present in the sample.

A flame test can be made for calcium by dissolving the white CaC_2O_4 precipitate in strong HCl solution, dipping a clean platinum wire into the solution, and placing it in a flame. The resulting brick-red or orange-red colored flame is not sufficiently characteristic to be relied on by inexperienced students. At any rate, the calcium oxalate precipitate is an excellent indication of the presence of calcium, and no other test is necessary.

EQUATIONS FOR REACTIONS IN STEP 48. The first reaction is the neutralization of nitric acid with ammonia solution. Nitric acid furnishes hydronium ions which are the active agent:

$$NH_3 + H_3O^+ \rightarrow NH_4^+ + H_2O.$$

The other reaction is the combination of the ions to form calcium oxalate:

$$Ca^{+2} + C_2O_4^{-2} \rightarrow \underline{CaC_2O_4} \text{ (White)}.$$

If the flame test is made, the reaction of HCl with calcium oxalate, forming calcium ions, results. Calcium ions with chloride ions in the flame are more volatile than they are with oxalate ions:

$$CaC_2O_4 + 2H_3O^+ \rightarrow Ca^{+2} + H_2C_2O_4 + 2H_2O.$$

Review Questions

1. Write out Chart 5 from memory.
2. From Chart 5 write equations for all reactions in all steps of the analysis of Group IV.

3. For the following groups of ions, each group contained in a single solution as chlorides, outline the steps needed to separate and make the final test for each ion.
   a. Pb^{+2}, Sn^{+2}, Sr^{+2}
   b. As^{+3}, Ba^{+2}
   c. Cu^{+2}, Al^{+3}, Ca^{+2}
   d. Hg^{+2}, Zn^{+2}, Mn^{+2}, Ba^{+2}
   e. Bi^{+3}, Cr^{+3}, Ca^{+2}
   f. Cd^{+2}, Ag^{+}, Fe^{+3}, Ni^{+2}, Sr^{+2}
4. Why are ammonium ions, as ammonium salts, added to the solution before Group IV carbonates are precipitated in the separation of Group IV from Group V?
5. Choose any pair of ions from Group IV and select a reagent or reagents that will precipitate one ion without precipitating the other.

7

Analysis of Group V

Group V is not a major group of ions to be separated and identified. In fact some books include it in Group IV. The ions present after the first four groups are separated are Mg^{+2}, Na^{+}, K^{+}, and NH_4^{+}. Ammonium ions are certain to be present at this point in the procedure because they have been added in several different steps along the way. Sodium ions are probably present also, because they are a common impurity in many reagents added in previous procedures. Although the amount of sodium thus added as a contaminant will no doubt be small, a flame test for sodium is so sensitive that it will probably be positive, if made on the solution from Step 41. It is possible to separate magnesium ions from those of sodium and potassium and then test for the latter ions by forming insoluble substances containing them as precipitates, but such procedures are not necessary. The salts of sodium and potassium are somewhat volatile and, like strontium and barium, they impart characteristic colors to a gas flame. The only ion that must be tested for by precipitation is that of magnesium (see Chart 6).

Because ammonium ions have been added as a reagent in previous steps, it is not possible to make a reliable test for them in this group. Nevertheless, the test for ammonium ions is given in this group, as is the custom in most texts.

Step 49. Precipitation of Traces of Group IV Ions. The solution from Step 41 may contain rather large traces of Ca^{+2}, Sr^{+2}, and Ba^{+2}. Barium and strontium ions are converted to the insoluble sulfates, and calcium, whose sulfate is slightly soluble, is precipitated as the oxalate.

Add 1 drop of 0.3 M ammonium oxalate and 1 drop of 0.2 M ammonium sulfate to the solution from Step 41. Warm in hot water a few minutes, cool, and centrifuge. The precipitate should be small

CHART 6

Analysis of Group V, the Sodium Group

The solution from Step 41 or a solution known to contain only elements of Group V may contain any or all of the following:

Mg^{+2}, K^{+}, Na^{+}, and NH_4^{+}.

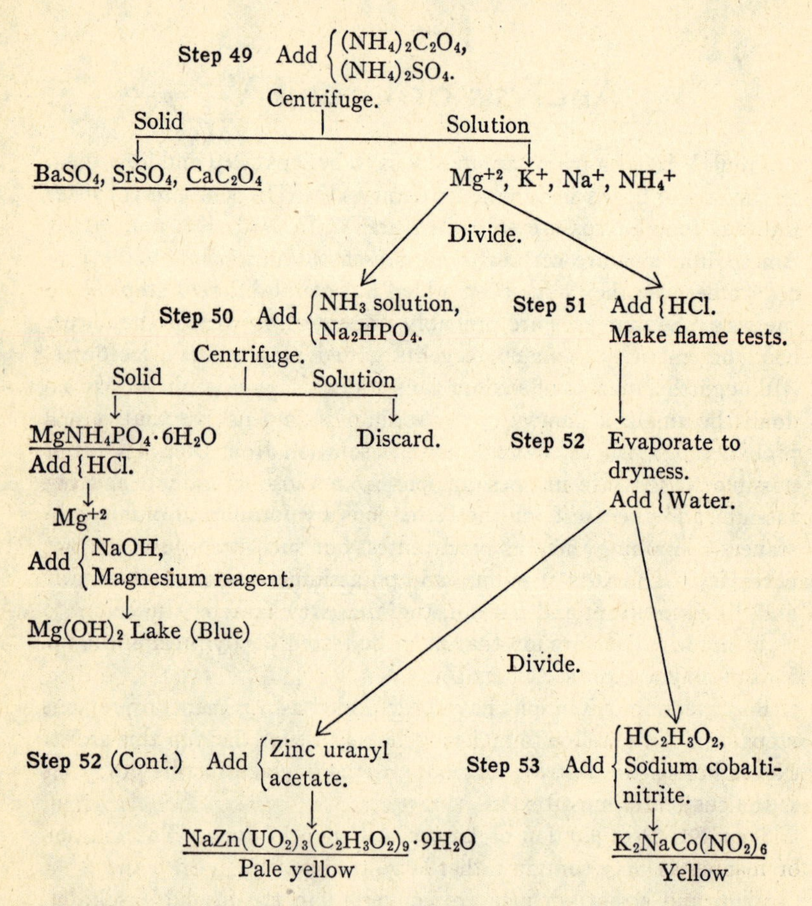

in bulk and consist of fine white crystals. Decant the solution into a small test tube. Discard the precipitate. The solution is ready for Step 50, the test for magnesium; Step 52, the precipitation test for sodium; and Step 53, the precipitation test for potassium. Flame tests should not be made on this solution because some contamination of the sample with sodium salts before this point is reached is unavoidable and because potassium salts may have been added during previous steps. Flame tests must be made on the original sample.

EQUATIONS FOR REACTIONS IN STEP 49. The equations in this step are for reactions in which ions combine to form insoluble salts:

$$Ba^{+2} + SO_4^{-2} \rightarrow \underline{BaSO_4} \text{ (White).}$$
$$Sr^{+2} + SO_4^{-2} \rightarrow \underline{SrSO_4} \text{ (White).}$$
$$Ca^{+2} + C_2O_4^{-2} \rightarrow \underline{CaC_2O_4} \text{ (White).}$$

Step 50. The Test for Magnesium. There are two tests for the detection of Mg^{+2}. The first is precipitation of white $MgNH_4PO_4 \cdot 6H_2O$ (magnesium ammonium phosphate hexahydrate) from a solution that is basic with ammonia. The second is a further test made by precipitating $Mg(OH)_2$ (magnesium hydroxide), which is white and very insoluble. $Mg(OH)_2$ is difficult to see, but it can be induced to adsorb certain dyes and become easily visible.

Withdraw about one-third of the solution from Step 49 and place it in a small test tube. Add 1 drop of 6 M NH_3 solution and 5 drops of 0.2 M Na_2HPO_4 (disodium phosphate) solution. Mix, heat, and allow to stand. If no precipitate forms in 3 to 4 minutes, Mg^{+2} was not present in the test solution. If a white precipitate forms, it is probably $MgNH_4PO_4 \cdot 6H_2O$, and it can be further confirmed in a test for magnesium as follows:

Centrifuge, discard the solution, and dissolve the precipitate with 1 drop of 6 M HCl and 2 or 3 drops of water. Add 2 drops of magnesium reagent, a dye in solution which turns blue when adsorbed on $Mg(OH)_2$, and 8 M NaOH solution a drop at a time, stirring with a glass rod. After each drop is added, remove the rod from the well-stirred solution without touching the walls of the test tube and touch it to litmus paper. After the test shows the solution to be alkaline, add 1 more drop. If a cloudy, bluish precipitate forms, the presence of magnesium is confirmed. If the test tube is given a whirl in a centrifuge, the blue precipitate can be easily seen in the bottom. The colored precipitate is a lake, a precipitated solid on which a dye is adsorbed. The dye in the magnesium reagent is *p*-nitrobenzeneazoresorcinol.

EQUATIONS FOR REACTIONS IN STEP 50. The first reaction is the precipitation of magnesium ammonium phosphate from a solution containing magnesium ions and ammonia, to which phosphate has been added:

$$Mg^{+2} + NH_4^+ + PO_4^{-3} + 6H_2O \rightarrow \underline{MgNH_4PO_4 \cdot 6H_2O} \text{ (White).}$$

The composition of the precipitate may vary from this formula somewhat. In the next reaction $MgNH_4PO_4 \cdot 6H_2O$ is dissolved in acid by the action of hydronium ions on phosphate ions, probably forming the dihydrogen phosphate ion and small amounts of phosphoric acid simultaneously. The equation for the reaction producing $H_2PO_4^-$ is:

$$2H_3O^+ + MgNH_4PO_4 \cdot 6H_2O \rightarrow Mg^{+2} + NH_4^+ + H_2PO_4^- + 8H_2O.$$

The final test for magnesium is the precipitation of magnesium hydroxide.

$$Mg^{+2} + 2OH^- \rightarrow \underline{Mg(OH)_2} \text{ (White + dye} \rightarrow \text{blue).}$$

The dye is probably adsorbed on the surface of the $Mg(OH)_2$. There is therefore no reaction between the precipitate and the dye.

Step 51. Flame Tests for Sodium and Potassium. Sodium and potassium salts, especially chlorides, are volatile, and the vapors impart a characteristic color to the flame of a burner. Therefore the most reliable means of detecting these two elements in compounds is by flame tests. Ammonium salts, if present, must be removed by sublimation before the tests are made, because they also are volatile and may impart to a flame a yellowish color which resembles that of sodium.

The sodium test is so sensitive that impurities from the glass of test tubes and beakers, from reagents used during previous procedures, or from merely touching a platinum wire with the fingers will give a positive test for sodium. Make all flame tests on the original sample before anything is done to it to contaminate it.

If the original sample is a solution, place 3 or 4 drops of it, or of the remainder of the solution from Step 49, in a crucible or casserole and gently evaporate to dryness. The solid that remains, or the original sample, if it is a solid, is heated in the casserole or crucible gradually until it is dull red. If a whitish smoke evolves during the heating, continue heating, increasing the heat gradually, until the white smoke of subliming ammonium salts ceases to be given off. At this point the heating may be stopped whether the crucible is red-hot or not.

After the ammonium salts are driven off, cool and add 2 drops of concentrated HCl and 1 drop of water. Clean a platinum wire by alternately dipping it in clean concentrated HCl and heating it in a flame until there is no longer a sodium or potassium flame color visible when the wire is inserted into the flame. Then dip the clean wire into the solution to be tested and insert the wire into a flame to the side of and above the central cone of the flame. All solutions give a slight sodium test. A positive test is an *intense* yellow color extending outward in all directions from the wire about $\frac{1}{8}$ to $\frac{1}{4}$ of an inch, with a column of yellow flame directly above the wire reaching to the top of the flame. The intensity of the flame will fade in a few seconds. A comparison should be made between the flame and that produced by a dilute solution of a sodium salt from the shelf.

The test for potassium is made on the same solution and at the same time as the test for sodium is made. Potassium ions impart a light violet color to a gas flame. The potassium flame is so much less intense than the sodium flame that it is hidden by the sodium flame unless one observes the test through a blue glass (cobalt or didymium). The blue glass filters out the yellow sodium flame color but permits the violet to pass through and be seen. The potassium flame is less sputtery than that of sodium. It begins to be visible rather quickly after the wire enters the flame and ends after only a second or two. The flame is smooth, rising straight up from the wire. A sample of a potassium salt from the shelf will acquaint the student with the character of the potassium flame. If the potassium flame is a mere flash, only a small amount of potassium is present. If the colored flame is large and continues a few seconds, potassium is present in larger amounts than traces.

The solution from Step 51 on which flame tests were made or the solution from Step 49 may be further tested for sodium and potassium by precipitation methods in Steps 52 and 53.

There are no equations for the reactions in Step 51.

Step 52. Precipitation Test for Sodium. Although the precipitation test for sodium is not very reliable, it illustrates the point that at least one sodium compound is only very slightly soluble. The insoluble compound produced is usually the pale-yellow sodium zinc uranyl acetate.

To make the test, evaporate the test solution from Step 49 or Step 51 to dryness in a casserole or crucible. Cool, add about 10 drops (0.5 ml) of water, and stir at least 1 full minute to dissolve the salts completely. Add 2 drops of this solution to 2 drops of zinc uranyl acetate solution (sometimes labeled "sodium reagent"). If a precipitate forms within several minutes, sodium ions were present in the test solution. If no precipitate forms, sodium might have been present in too small amounts to give a precipitate.

EQUATION FOR THE REACTION IN STEP 52. The only reaction is the precipitation of the sodium zinc uranyl acetate:

$$Na^+ + Zn(UO_2)_3(C_2H_3O_2)_9^- + 9H_2O \rightarrow \underline{NaZn(UO_2)_3(C_2H_3O_2)_9 \cdot 9H_2O} \text{ (Pale yellow).}$$

Step 53. Precipitation Test for Potassium. Although this test is somewhat more dependable than the precipitation test for sodium, it is not altogether reliable. Ammonium salts react exactly like potassium in forming a similar precipitate, and considerable amounts of potassium ion may be present without a precipitate being formed.

Add 2 drops of the solution prepared for the sodium test in Step 52 to a test tube. Add 1 drop of 5 M acetic acid and 3 drops of sodium cobaltinitrite (potassium reagent) solution. Stir and allow to stand. If a yellow precipitate forms within 5 minutes, the presence of potassium is indicated. If no precipitate forms, potassium may be present but not in large quantities.

EQUATION FOR THE REACTION IN STEP 53. The proper name for sodium cobaltinitrite (potassium reagent) is trisodium hexanitritocobaltate(III). This reagent dissolves to form sodium ions and hexanitritocobaltate(III) ions. If it is added to a solution which contains potassium ions, sodium, potassium, and hexanitritocobaltate(III) ions combine to form dipotassium sodium hexanitritocobaltate(III):

$$2K^+ + Na^+ + Co(NO_2)_6^{-3} \rightarrow \underline{K_2NaCo(NO_2)_6} \text{ (Yellow).}$$

Step 54. The Detection of Ammonium Ions. This test is always made on original sample material. Ammonium salts are used as reagents and will give positive tests for ammonium ion (NH_4^+) whether it is present in the original sample or added later. If the sample is a solution, it is ready for testing. If the sample is a solid, place a small amount of the solid, about 0.1 g, in the bottom of the smallest beaker available. Add 1 drop of water and stir the solid with the water for 1 minute if not all of the solid dissolves. The material is then ready for testing.

To a solution prepared from the solid sample as described above or to 2 drops of the sample solution in the bottom of a very small beaker (D in Fig. 6), add a pellet of solid sodium hydroxide, NaOH. The

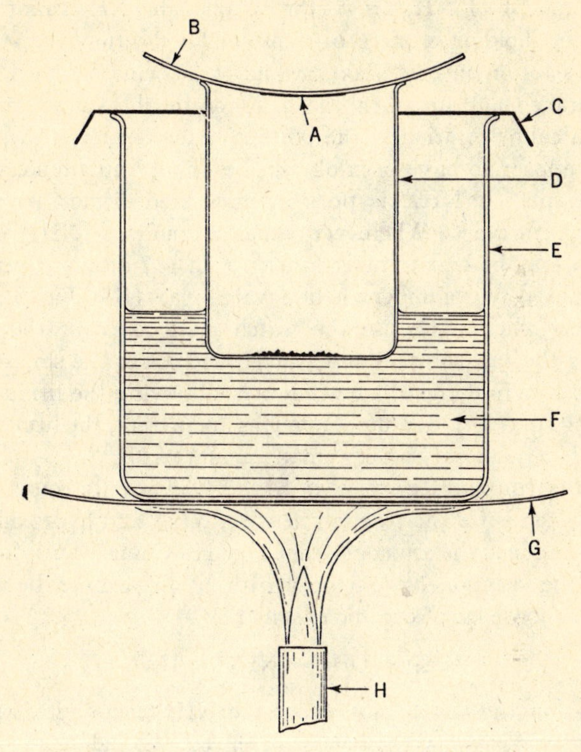

Fig. 6. The Test for Ammonia. A is red litmus paper moistened and placed on the underside of the watch glass, B. C is a cover for the larger beaker, E, with a hole in it just large enough to allow beaker D to slip down through. Beaker D contains the sample and some sodium hydroxide solution. F is hot water, G is a wire screen, and H is a burner.

procedure from here on varies from book to book. All authors agree the mixture should be warmed. NaOH converts NH_4^+ to NH_3, which is very soluble in water at low temperatures but not soluble at near-boiling temperatures. Heating, therefore, drives off NH_3 as vapor, but the heating must be done with caution. If the solution is boiled, the strong NaOH solution will spatter, causing litmus paper to turn blue without any ammonia being present. It is best to heat the little beaker in a larger one containing water (see Fig. 6).

The detection of NH_3 in a sample may be accomplished by smelling the vapors *if* a rather large amount of NH_4^+ is in the sample. If only a small amount of NH_4^+ is in the sample, the NH_3 vapor can be detected by holding a piece of moistened red litmus paper down near the solution in the beaker while it is being warmed. Care must be taken not to touch the sides of the beaker or the very strongly alkaline solution in the bottom of the beaker, or a positive test may appear to have been obtained without any ammonium ion being present. If NH_4^+ is present, moist red litmus paper will gradually turn blue. Where very small amounts of NH_4^+ may be present, it may be best to moisten a strip of red litmus paper (A in Fig. 6), stick it to the underside of a watch glass (B in Fig. 6) (it will adhere well when wet), use the watch glass to cover the beaker containing the test solution and NaOH, and warm the small beaker in a water bath (see Fig. 6) a few minutes, observing the litmus paper for a change in color. If ammonium ions are present, the litmus paper will turn blue with only a very small amount of NH_4^+.

Equation for the Reaction in Step 54. The only equation that can be written is for the reaction between NH_4^+ and hydroxide from NaOH to produce the ammonia that changes litmus paper from red to blue (the reaction of ammonia with litmus cannot be written without a knowledge of organic chemistry):

$$NH_4^+ + OH^- \rightarrow NH_3\uparrow + H_2O.$$

Of course, on the wet litmus paper the NH_3 reacts with water to produce hydroxide ions, the reverse of the equation above.

Review Questions

1. Write out Chart 6 from memory.
2. Write equations for all reactions occurring in the Group V analysis.
3. If traces of barium, strontium, and calcium ions are not completely removed in the Group IV analysis, what one would not be completely removed by sulfate only in Step 49? Which ion is most completely precipitated by sulfate in Step 49?
4. Why is it very important not to boil a solution being tested for ammonium ions after sodium hydroxide has been added?
5. Why should one make sodium and potassium flame tests on the original sample rather than on the solution after magnesium is removed?

6. Why would the test for ammonium ions made on the solution from Step 49 be unreliable?
7. What reagent will give a precipitate with a solution of the following?
   a. $(NH_4)_2SO_4$ but not with Na_2SO_4
   b. KCl but not NaCl
   c. NaCl but not KCl

8

Analysis of a Metal or Alloy

After completing the analysis of all five groups of metal ions, a student is able to analyze a large number of alloys. These include the brasses, bronzes, bearing metals, casting alloys, and solders. Some steels can also be analyzed, although many contain elements such as tungsten, molybdenum, and vanadium which are not included in the procedures given in elementary books. In addition to metals, many nonmetals such as sulfur, phosphorus, and silicon may be found in alloys, sometimes as impurities. These nonmetals can be detected by treating the alloy with acids and applying the methods outlined in Chapter 9. Sulfur as sulfide may be converted to H_2S by HCl or H_2SO_4 treatment. It may also be converted to sulfate with HNO_3 and detected as such. Phosphorus may be converted to phosphate with HNO_3, and silicon may be converted to $SiO_2(H_2O)_x$ with HCl or $HClO_4$ (perchloric acid).

Some alloys dissolve only slowly or not at all in nitric acid or in nitric-hydrochloric acid (aqua regia) solutions. Some alloys require fusion with reagents such as potassium pyrosulfate, $K_2S_2O_8$, or sodium carbonate, Na_2CO_3, plus NaOH, followed by acid treatment. Almost all alloys dissolve in perchloric acid, but special hoods are required when it is used, because perchloric acid and its fumes form violently explosive mixtures with combustible organic materials such as are found in dust or laboratory vapors.

The only alloys considered here are those which are attacked by 1–1 HNO_3 (1 part water, 1 part concentrated HNO_3) or HNO_3–HCl solution (aqua regia). The general procedures are outlined in Chart 7.

Step 55. Dissolving a Metallic Sample. If a balance is available, weigh out 0.03 to 0.05 g of sample into a small beaker. If no balance is available, the instructor can display a small test tube containing a sample weighed out for comparison. Add 4 drops of water and 4 drops

CHART 7

ANALYSIS OF A METAL OR ALLOY (NITRIC-ACID-SOLUBLE)

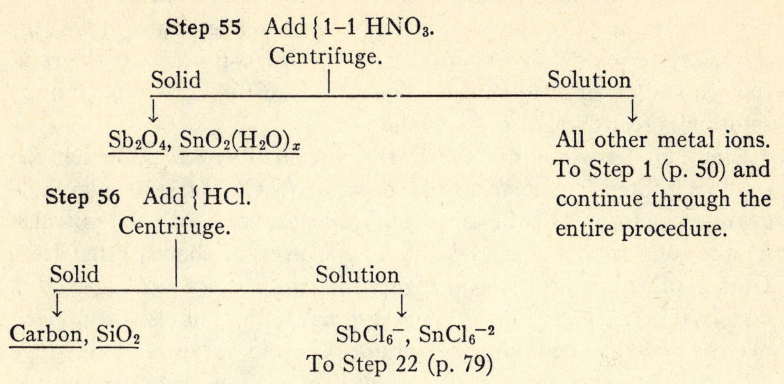

METAL OR ALLOY (NOT NITRIC-ACID-SOLUBLE)

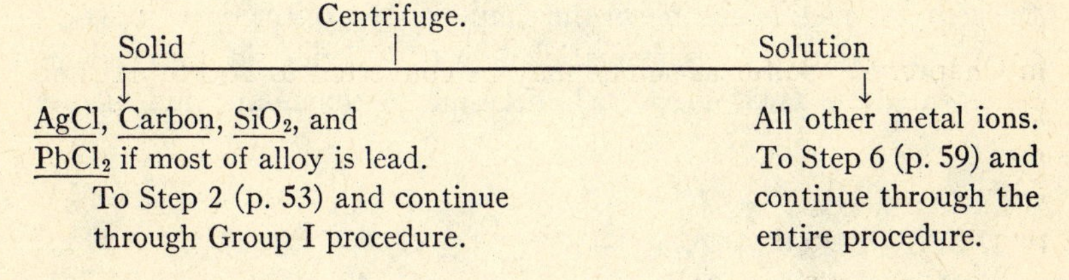

of concentrated nitric acid and cover with a watch glass. If a reaction starts immediately, do not heat. If no reaction is apparent, warm gently at first, then more vigorously until the solution boils. Do not boil at length or the solution will all evaporate. If there still is no reaction and if no bubbles continue to rise from the metal particles after the liquid in the beaker is cooled below boiling, the alloy is insoluble in HNO_3 only and must be treated as in Step 57. If bubbles do continue to rise from the particles after the acid is cooled below boiling, a reaction is occurring. Continue heating at just below boiling until the alloy completely deteriorates or dissolves. This may require 5 to 10 minutes, and more water and acid in equal parts must be added to maintain about 0.5 to 1.0 ml of volume. Darkish, heavy chunks left in the bottom of the beaker indicate incomplete action. A small, black, fluffy residue which does not seem to react after 10 minutes of treatment may be carbon or other nonmetallic substances, especially if the alloy is a steel or cast iron. A white residue of finely divided solid indicates that the alloy contained tin or antimony; the white material is either hydrated stannic oxide, $SnO_2(H_2O)_x$, or antimony tetroxide, Sb_2O_4, or both. Centrifuge and decant the solution into a beaker or test tube. The solid is ready for Step 56 and the solution is ready for Step 1 (p. 50).

Step 56. Treatment of the Residue from Step 55. Whether the solid from Step 55 is black, gray, or white (a black residue may hide a white one formed at the same time), wash it once with water, transfer the solid to a beaker and add 3 to 5 drops of concentrated HCl. Heat gently, but do not boil vigorously, until the white material is dissolved or, if the residue is black or gray, for 5 minutes. More HCl may be added to maintain the volume of liquid uniform. Centrifuge if there is any undissolved material, discard the solid, and the solution is ready for Step 22 (p. 79).

EQUATIONS FOR REACTIONS IN STEPS 55 AND 56. In alloys most metals act chemically as if they were alone, even though they may sometimes exist as compounds in the alloy. The equations are for the common metals in some of the simple alloys, in the order in which they appear during the systematic analysis. Nitric acid in solution is represented by $H_3O^+ + NO_3^-$, hydronium (or oxonium) and nitrate ions:

1. $3Ag + 4H_3O^+ + NO_3^- \rightarrow 3Ag^+ + NO\uparrow + 6H_2O$.
2. $3Pb + 8H_3O^+ + 2NO_3^- \rightarrow 3Pb^{+2} + 2NO\uparrow + 12H_2O$.
3. $3Hg + 8H_3O^+ + 2NO_3^- \rightarrow 3Hg^{+2} + 2NO\uparrow + 12H_2O$.
4. $Bi + 4H_3O^+ + NO_3^- \rightarrow Bi^{+3} + NO\uparrow + 6H_2O$.
5. $3Cu + 8H_3O^+ + 2NO_3^- \rightarrow 3Cu^{+2} + 2NO\uparrow + 12H_2O$.
6. $3Cd + 8H_3O^+ + 2NO_3^- \rightarrow 3Cd^{+2} + 2NO\uparrow + 12H_2O$.
7. $3Sn + 4H_3O^+ + 4NO_3^- \rightarrow 3SnO_2 + 4NO\uparrow + 6H_2O$.
8. $3As + 6H_2O + 5NO_3^- \rightarrow 3AsO_4^{-3} + 5NO\uparrow + 4H_3O^+$.
9. $6Sb + 10H_3O^+ + 10NO_3^- \rightarrow 3Sb_2O_5 + 10NO\uparrow + 15H_2O$.

Hydrated Sb_2O_5 is first formed (see p. 11), but at the temperature of the boiling solution it is partially decomposed to Sb_2O_3:

$$Sb_2O_5 \rightarrow Sb_2O_3 + O_2.$$

Sb_2O_3 then combines with the remaining Sb_2O_5 to form the combined oxide Sb_2O_4, in which one Sb atom has an oxidation state of +3 and the other +5:

$$Sb_2O_3 + Sb_2O_5 \rightarrow 2Sb_2O_4.$$

With nitric acid, Ni, Co, Mn, Zn, Ca, Ba, Sr, and Mg react to form divalent ions like Pb, Hg, Cu, and Cd. The equations for nitric acid reacting with them are analogous to equations 2, 3, 5, and 6 above.

Although Fe, Al, and Cr exhibit what is known as "passivity" to concentrated HNO_3, they do dissolve in dilute HNO_3 solutions. Equations for the reactions are:

10. $Fe + 4H_3O^+ + NO_3^- \rightarrow Fe^{+3} + NO\uparrow + 6H_2O.$
11. $Al + 4H_3O^+ + NO_3^- \rightarrow Al^{+3} + NO\uparrow + 6H_2O.$
12. $Cr + 4H_3O^+ + NO_3^- \rightarrow Cr^{+3} + NO\uparrow + 6H_2O.$

Sodium and potassium are rare in alloys and are such active metals that acids are not needed to dissolve them. They react violently with water to form sodium hydroxide and hydrogen:

13. $2Na + 2H_2O \rightarrow 2Na^+ + H_2\uparrow + 2OH^-.$
14. $2K + 2H_2O \rightarrow 2K^+ + H_2\uparrow + 2OH^-.$

The equations for reactions of nitric acid with metals all indicate that nitric oxide, NO, is the reduction product of nitric acid. It should be noted that nitrogen dioxide, NO_2, is also produced and in some cases may be the main reduction product of nitric acid. All the equations given above can therefore be written with NO_2 instead of NO as a product. For example, Equation 1 can be written:

$$Ag + 2H_3O^+ + NO_3^- \rightarrow Ag^+ + NO_2\uparrow + 3H_2O,$$

and Equation 2:

$$Pb + 4H_3O^+ + 2NO_3^- \rightarrow Pb^{+2} + 2NO_2\uparrow + 6H_2O.$$

Equation 8 is changed to:

$$As + 2H_3O^+ + 5NO_3^- \rightarrow AsO_4^{-3} + 5NO_2\uparrow + 3H_2O.$$

Step 57. Treatment of a Sample Insoluble in Nitric Acid. If the alloy is not attacked by nitric acid alone after 5 minutes, add, a drop at a time, a volume of concentrated HCl equal in volume to the HNO_3 solution remaining. If action is slow, warm to near boiling. Treatment with both nitric and hydrochloric acids dissolves tin and antimony and all metals but silver and — if a large amount is present — lead. Silver and lead chloride and undissolved SiO_2 will remain as a white residue if silver, lead, or silicon is present in the alloy. Lead chloride is very soluble in hot water and may not appear at this point unless the solution is cooled. After the alloy is completely disintegrated, if there is a residue, add 10 drops of distilled water, centrifuge, and the residue will be ready for Step 2 (p. 53). The solution is ready for Step 6 (p. 59). No Hg_2Cl_2 will be formed, because HNO_3 will oxidize mercury to Hg^{+2}, which will remain in solution as $HgCl_2$ or $HgCl_4^{-2}$.

A black residue may be carbon, which may hide some white solid. Treat it like a white residue until assurance is obtained that it does not hide AgCl or $PbCl_2$.

EQUATIONS FOR REACTIONS IN STEP 57. In Step 57 alloys not soluble in nitric acid alone are treated with a mixture of nitric and hydrochloric acids.

Chromium, aluminum, and iron alloys do not always dissolve in nitric acid solutions. This is thought to be due to the fact that the oxidizing action of nitric acid forms an inert oxide coating over the surface and thus protects the alloy from further action by reagents. The use of HCl usually breaks down this oxide coating so that the alloys dissolve in the solution of both nitric and hydrochloric acids.

Note that an excess of chloride ions is needed for formation of chloride complex ions in equations 16 through 21. The strong HCl solution furnishes ample chloride in excess. The formation of the stable complexes helps make these reactions go to completion:

15. $3Ag + NO_3^- + 3Cl^- + 4H_3O^+ \rightarrow \underline{3AgCl} + NO\uparrow + 6H_2O.$
16. $3Hg + 2NO_3^- + 12Cl^- + 8H_3O^+ \rightarrow 3HgCl_4^{-2} + 2NO\uparrow + 12H_2O.$
17. $3Sn + 4NO_3^- + 18Cl^- + 16H_3O^+ \rightarrow 3SnCl_6^{-2} + 4NO\uparrow + 24H_2O.$
18. $3Sb + 5NO_3^- + 20H_3O^+ + 18Cl^- \rightarrow 3SbCl_6^- + 5NO\uparrow + 30H_2O.$
19. $Fe + NO_3^- + 4H_3O^+ + 6Cl^- \rightarrow FeCl_6^{-3} + NO\uparrow + 6H_2O.$
20. $Al + NO_3^- + 2H_3O^+ + 4Cl^- \rightarrow AlCl_4^- + NO\uparrow + 3H_2O.$
21. $Cr + NO_3^- + 4H_3O^+ + 6Cl^- \rightarrow CrCl_6^{-3} + NO\uparrow + 6H_2O.$

In Step 57 the other metals react to produce the same products as in Steps 55 and 56. The equations for the reactions are therefore the same as those given for Steps 55 and 56.

As in the reactions in Steps 55 and 56, the reactions in Step 57 may yield NO_2 as one product as well as NO. The equations above can be modified to indicate NO_2 as the reduction product of nitric acid in the same way as is described on p. 123.

Review Questions

1. Make an outline of the steps to be taken to analyze an alloy.
2. Choose any common alloy such as brass, bronze, solder, or bearing metal. Find what it consists of and trace each element through the entire analysis of the alloy and the ions. Make a chart showing each operation needed to dissolve the alloy, separate the ions, and test for each. Do this for all of the common alloys.
3. Write out equations for all reactions employed in your charts in question 2.
4. A bearing metal is suspected of containing silver. If the other elements in the alloy are lead, tin, and antimony, what steps would be needed to show the presence or absence of silver?

9

Analysis of Salts

To analyze salts or mixtures of salts one must apply procedures for the cation analysis, and in addition one must detect all anions (negative ions). Although there is no orderly procedure for separating and identifying the anions, as there is for the cations, definite steps must be taken to test for or prove the absence of each ion. In almost all cases several ions can be proved absent by a single test, thereby eliminating the need to make individual tests for those ions eliminated. Furthermore, there are not many common anions and most of them can be identified in the presence of other anions; thus elaborate procedures for separating the ions from each other are unnecessary.

The common anions can be divided into three groups:

1. One group, called the *chloride group*, forms salts with silver ions that are insoluble in dilute nitric acid. The ions of the chloride group selected for analysis in this book are chloride (Cl^-), bromide (Br^-), iodide (I^-), and sulfide (S^{-2}).

2. A second group of anions, called the *sulfate group*, forms salts with barium or calcium ions that, except for $BaSO_4$, are soluble in dilute nitric acid but are insoluble in water. The ions selected from this group for analysis are sulfate (SO_4^{-2}), sulfite (SO_3^{-2}), phosphate (PO_4^{-3}), arsenate (AsO_4^{-3}), borate (BO_2^-), carbonate (CO_3^{-2}), chromate (CrO_4^{-2}), oxalate ($C_2O_4^{-2}$), and fluoride (F^-).

3. The anions of a third group do not form insoluble salts with either silver, barium, or calcium ions. These are acetate ($C_2H_3O_2^-$), nitrate (NO_3^-), and nitrite (NO_2^-), and because the nitrate ion is the most common of the ions of this, the soluble group, it is often called the *nitrate group*.

Analysis of a Sample

Before beginning the analysis of a salt or mixture of salts it is necessary to have a plan or outline of steps to take and the proper order in which to take them. The plan or outline will vary with the type of sample to be analyzed.

If the sample to be analyzed is a solution:

1. First analyze half of the sample for the anions beginning with Step 63 (p. 135).

2. Analyze the other half of the sample for the cations in the following order:

a. Use a small portion to test for the ammonium ion (NH_4^+), by Step 54 (p. 116).

b. Test for K^- and Na^- by the flame test, Step 51 (p. 114).

c. Analyze the remaining solution, starting with Step 1 (p. 50).

If the sample to be analyzed is a solid, use a small portion for the test for the ammonium ion in Step 54 (p. 116). Use another small portion for the sodium and potassium flame tests, Step 51 (p. 114); then proceed with Step 58 and the complete anion analysis, using a portion of the sample. Save a portion of the sample for individual tests and for the cation analysis. Upon completion of the anion analysis, proceed with the complete analysis for the cations, beginning with Step 1 (p. 50), using either the water solution of the sample or, if the sample is not water soluble, the nitric acid solution of the precipitate obtained in Step 62 (p. 133), in which the sample is treated with sodium carbonate.

Step 58. Inspection of the Sample. If the sample is colored, the color reveals something of the nature of the ions in it. If the sample is colorless, all colored ions are absent and need not be tested for. Certain ions are colorless when dry, like the copper ion in anhydrous copper sulfate, but acquire their characteristic color when water is added. Inspect the sample to see if it is homogeneous and identify any and all colors. Then consult Table 2, p. 128, to see what possible substances may be suggested by the colors you saw in the sample. Do not rely entirely on colors seen for identification of ions and compounds. Always confirm your guesses with proper chemical tests.

Step 59. The Test for Solubility of the Sample in Water. Place a few very small grains or some of the powdered sample in a test tube. The total quantity of solid taken should not exceed one-fourth the size of a grain of rice or wheat. Add 2 drops of water and stir for 25

TABLE 2

Color	Dry Solid	Ion in Solution
Black	Ag_2S, PbS, CuS, HgS, CoS, NiS, FeS, CuO, Fe_3O_4, MnO_2, NiO	
Yellow	As_2S_3, As_2S_5, SnS_2, CdS	CrO_4^{-2}, Fe^{+3}
Orange	Sb_2S_3, dichromates ($Cr_2O_7^{-2}$), and sometimes CdS	$Cr_2O_7^{-2}$, Fe^{+3}
Blue	Cu^{+2}, Ni^{+2}, and Fe^{+2} salts and anhydrous Co^{+2} salts	Cu^{+2}, Ni^{+2}, Fe^{+2}
Green	Cu^{+2}, Ni^{+2}, Fe^{+2}, and some Cr^{+3} salts	Cr^{+3}, Cu^{+2}, Ni^{+2}, Fe^{+2}
Purple	MnO_4^- (permanganate) and some Cr^{+3} salts	MnO_4^-
Red	HgS, Sb_2S_3, HgO, Pb_3O_4, Fe_2O_3, and sometimes As_2S_3	$FeSCN^{+2}$ (ferric thiocyanate)
Brown	Bi_2S_3, SnS, Ag_2O, Bi_2O_3, CdO, PbO_2, $CuCrO_4$, $Fe_2O_3(H_2O)_x$	

to 30 seconds. If the solid dissolves completely, the sample is *soluble* in water. If the solid does not dissolve in 2 drops of water, add 5 more drops of water and stir for another 30 seconds. If the salt now dissolves, it is *somewhat soluble* in water. If it does not dissolve, add 10 more drops of water (17 drops in all) and stir again. If the sample dissolves now, it is *sparingly soluble* in water. If the sample does not dissolve in 17 drops of water, it is *very slightly soluble* or *insoluble*.

If all or part of the sample dissolved in water, dissolve a larger portion; then divide the solution into two parts. One part is ready for Step 63A (p. 135) and the anion analysis, while the other part is ready for Step 54 (p. 116), Step 51 (p. 114), and then Step 1 (p. 50) and the entire cation analysis.

That part of the sample tested in this step which was insoluble in water is ready for Step 60.

Step 60. The Test for Solubility in Hydrochloric Acid. Remove all liquid from the tube in which the undissolved solid remains from Step 59. Now add 1 drop of 6 M HCl solution by allowing the acid to run down the side of the tube. Watch the solid closely when the acid contacts it. If the drop of acid does not cause any visible reaction — that is, if no bubbles of gas form or if no change in the appearance of the solid can be seen — add 3 more drops of 6 M HCl solution and stir for 30 seconds. If the solid still does not show signs of dissolving, add 6 or 8 drops of distilled water and stir another 30 seconds. If the solid does not dissolve after this treatment, it can be considered insoluble in dilute HCl, and a portion of the original sample must be treated as described in Step 61 (p. 130).

If the solid does dissolve in HCl, add about 0.2 g of sample to the test solution, dissolve it in HCl, and treat the solution as described in Step 6 (p. 59) and all the following steps in the complete cation analysis. A portion of the sample must also be treated as described in Step 61, whether the sample was soluble in HCl solution or not.

Besides very important solubility information, Steps 58, 59, and 60 eliminate a great many ions from among those that could be present in a sample. The solubility product table in Appendix VI lists the substances usually found in samples that are insoluble in water. A few facts about solubilities in general are helpful if they are remembered and applied. Three of them are:

1. Carbonates, sulfites, borates, phosphates, arsenates, chromates, and fluorides of most metal ions except sodium, potassium, and ammonium are insoluble in water but soluble in dilute HCl. However, fluorides form complex ions with ferric and aluminum ions that are soluble and very stable. Phosphate also forms a soluble colorless stable complex with ferric ions in acid solution.

2. Sulfides of copper and mercuric mercury (Hg^{+2}) are insoluble in dilute HCl solution. Nickel and cobalt sulfides do not dissolve readily in dilute HCl, a nonoxidizing acid, unless they have been freshly precipitated. CaS, BaS, SrS, K_2S, and Na_2S are soluble in water, but, like Al_2S_3, they hydrolyze (see p. 28) to give solutions that smell of H_2S. Sulfides of other metals are generally soluble in HCl solution but not in water. Many sulfides are colored, which aids in their identification.

The hydrolysis (see p. 42) of sulfides such as CaS and Al_2S_3 occurs because these sulfides are somewhat soluble in water and their ions react with water to form either slightly ionized, volatile, or insoluble products. CaS reacts to produce slightly ionized and volatile H_2S:

$$CaS + H_2O \rightarrow Ca(OH)_2 + H_2S \uparrow .$$

Al_2S_3 reacts to produce slightly ionized, volatile H_2S and very insoluble $Al(OH)_3$:

$$Al_2S_3 + 6H_2O \rightarrow \underline{2Al(OH)_3} + 3H_2S \uparrow .$$

3. Some judgment is required in deciding whether a substance is soluble in water or not, or soluble in HCl solution or not. Large amounts of solid obviously require large amounts of solvent, and sparingly soluble substances are often reported by students as being insoluble because the test was not made with care. Start tests with a small amount of sample, and a small amount of solvent. Gradually increase the amount of solvent until the substance dissolves or until the volume of solvent is at least 50 times that of the solid whose solubility is being tested. It is important to know if a substance is very soluble, somewhat soluble, sparingly soluble, or insoluble.

One other point regarding solubilities. Bismuth, antimony, and other elements form compounds that have a tendency to react with water by protolysis, often called "hydrolysis" (see pp. 28, 38, and 129), to produce insoluble solids. The hydrolysis of bismuth and antimony chlorides can be illustrated by the following equations:

$$Bi^{+3} + Cl^- + 3H_2O \rightleftharpoons 2H_3O^+ + \underline{BiOCl}\text{ (White).}$$
$$Sb^{+3} + Cl^- + 3H_2O \rightleftharpoons 2H_3O^+ + \underline{SbOCl}\text{ (White).}$$

The addition of acid reverses these reactions. Hydrolysis of these ions is recognized by the formation of a powdery solid in the water solution. Try some antimony chloride ($SbCl_3$) from the shelf in water in order to become familiar with the appearance of an insoluble hydrolysis product.

Step 61. The Preliminary Test with Sulfuric Acid. This test is performed only on solids, and only if some assurance from the instructor is available that no permanganate or chlorate ions are present in the sample. It is a test which indicates a great deal about the nature of the anions in the sample. It is of very little value in indicating anything about the cations.

To a small bit (about one-fourth the size of a grain of rice or wheat) of solid sample in the bottom of a test tube add 1 drop of concentrated sulfuric acid, allowing it to run down the inside wall of the test tube. Watch closely for any sign of a reaction when the acid contacts the sample. Look for the formation of a product which is either a gas or a substance different in color from the original sample. Both things may happen at once. Do not let one change cause you to overlook the other. If no reaction is evident after a few seconds, warm the test tube gently, pointing the top of it toward a hood or some other inanimate object. Neither you nor your neighbor cares to be spattered with hot sulfuric acid from a test tube heated too rapidly or in which a violent reaction is generated by heat.

Consult the following list of ions and how they react when tested with sulfuric acid, and decide what ions are probably absent, what ones are probably present, and what ones are certainly not present in the sample. Of course, those certainly not in the sample need not be tested for individually in Step 65 and other anion tests.

1. Arsenates, borates, phosphates, and (of course) sulfates show no reaction with H_2SO_4, cold or hot.

2. Acetate and nitrate do not react visibly with cold H_2SO_4, but warming causes acetate to yield volatile acetic acid with the odor of vinegar. The equation is:

$$C_2H_3O_2^- + H_2SO_4 \rightarrow HC_2H_3O_2\uparrow + HSO_4^-.$$

Nitrate reacts in two steps: first, HNO_3 is produced, and then the HNO_3 is decomposed by heat to produce NO_2, a brown gas. Equations for the reactions are:

$$NO_3^- + H_2SO_4 \rightarrow HNO_3 + HSO_4^-.$$

$$4HNO_3 \xrightarrow{\Delta} O_2\uparrow + 2H_2O + 4NO_2\uparrow.$$

3. Although yellow chromate (CrO_4^{-2}) does not undergo a very spectacular change, it is converted into dichromate, which is orange or red:

$$\text{Yellow } 2CrO_4^{-2} + H_2SO_4 \rightarrow H_2O + SO_4^{-2} + Cr_2O_7^{-2} \text{ (Orange)}.$$

4. Chloride and fluoride react alike with sulfuric acid, to some degree. Cold sulfuric acid produces HCl with chlorides and HF with most fluorides. Both HCl and HF are colorless; both have a sharp odor and should not be inhaled — HF is extremely poisonous. Moist air blown gently across the top of the test tube will produce a fog with either HCl or HF. Blow the air slowly, with your mouth fairly wide open and several inches from the top of the test tube. Do not breathe in without removing your face from the neighborhood of the top of the test tube. The equations for the reactions forming HCl and HF are:

$$Cl^- + H_2SO_4 \rightarrow HSO_4^- + HCl\uparrow.$$

$$F^- + H_2SO_4 \rightarrow HSO_4^- + HF\uparrow.$$

Bromide and iodide produce the same type of fog with moist breath, but sulfuric acid also produces the free elements with these ions and the elements are colored.

5. Bromide with cold sulfuric acid reacts like chloride:

$$Br^- + H_2SO_4 \rightarrow HSO_4^- + HBr\uparrow.$$

But bromide is a reducing agent and is oxidized by sulfuric acid to free bromine, so that the HBr is discolored yellow or red-brown by the free element. The reaction to yield free bromine may be indicated by the equation:

$$2HBr + H_2SO_4 \rightarrow SO_2\uparrow + 2H_2O + Br_2\uparrow.$$

6. Iodide reacts instantly with cold sulfuric acid to produce free iodine, which has a very dark purple or black color. At the same time HI, which produces a fog with moist air, is given off, mixed with violet iodine (I_2) vapors and either sulfur dioxide or hydrogen sulfide, or both. The latter two gases can be detected by odor. SO_2 smells like burning sulfur, and H_2S smells like rotten eggs. I_2 vapors, SO_2, and H_2S are poisonous gases. Sniff them cautiously.

$$H_2SO_4 + I^- \rightarrow HSO_4^- + HI\uparrow.$$
$$2HI + H_2SO_4 \rightarrow SO_2\uparrow + 2H_2O + I_2\uparrow.$$
$$8HI + H_2SO_4 \rightarrow H_2S\uparrow + 4H_2O + 4I_2\uparrow.$$

7. Carbonate (CO_3^{-2}) and sulfite (SO_3^{-2}) produce bubbles of colorless gases rapidly with cold sulfuric acid. Carbonate yields CO_2, which has no distinctive odor. Sulfite produces SO_2, which has the sharp characteristic odor of burning sulfur. The equations for the reactions are:

$$CO_3^{-2} + 2H_2SO_4 \rightarrow 2HSO_4^- + CO_2\uparrow + H_2O.$$
$$SO_3^{-2} + 2H_2SO_4 \rightarrow 2HSO_4^- + SO_2\uparrow + H_2O.$$

8. Sulfide (S^{-2}) reacts with cold sulfuric acid more or less rapidly depending on the solubility of the particular sulfide. Very insoluble sulfides hardly react at all unless heated. More soluble sulfides such as ZnS react vigorously with cold sulfuric acid, producing H_2S as well as some other products such as sulfur and SO_2. SO_2 is produced as a result of reduction of sulfuric acid by H_2S. Equations for possible reactions that will yield these products are:

$$2H_2SO_4 + S^{-2} \rightarrow 2HSO_4^- + H_2S\uparrow.$$
$$H_2SO_4 + H_2S \rightarrow 2H_2O + SO_2\uparrow + S.$$

9. Oxalate ($C_2O_4^{-2}$) reacts with cold H_2SO_4 to produce CO_2, a gas, but the reaction is not very spectacular. If warmed, the reaction is

faster and some charring may take place. Equations for reactions which will produce CO_2 are:

$$2H_2SO_4 + C_2O_4^{-2} \rightarrow H_2C_2O_4 + 2HSO_4^-.$$
$$H_2C_2O_4 \rightarrow CO_2\uparrow + CO\uparrow + H_2O.$$

$H_2C_2O_4$ is oxalic acid, from which concentrated sulfuric acid can withdraw water, leaving CO_2 and CO.

10. Oxides of all metals except sodium and potassium are insoluble in water but will react to some degree with cold sulfuric acid and more rapidly when hot, to produce sulfates. The active metal oxides such as CaO, SrO, and BaO may react very vigorously with sulfuric acid and may produce insoluble sulfates. Most oxides react more slowly and produce sulfates that are soluble in water.

Step 62. Preparation of the Anion Test Solution. Whether the sample is soluble or insoluble in water or acid, so long as it is a solid salt or mixture of salts, a portion of the sample must be treated with sodium carbonate solution to form hydroxides or carbonates of all cations and to form sodium salts of all anions so that most of the anions can be tested for individually if necessary.

Place an amount of sample the size of 2 or 3 grains of rice in a casserole or beaker. Add 4 or 5 ml of saturated sodium carbonate solution and heat to boiling. Stir and maintain at the boiling point for 2 or 3 minutes. If all lumps in the solution have not disintegrated into powder, heat and stir longer. Centrifuge and decant the clear solution into a flask or bottle that can be stoppered, and save it as "the anion test solution." It is ready for Step 63B and subsequent steps in the anion analysis.

The solid left from centrifuging consists of carbonates or hydroxides, or both, of all metals except sodium and potassium. If the original sample was soluble in water or HCl solution discard this solid, but if the sample was not soluble in water or HCl solutic proceed as follows:

To the solid in the test tube add, a drop at a time, 3 M HNO_3 solution until carbon dioxide ceases to be produced. Then add concentrated nitric acid a drop at a time, stirring and warming in boiling water between additions of acid until all of the solid is dissolved. This solution is now ready for Step 1 and all the following steps in the cation analysis. Obviously, if the original sample was soluble in water or HCl solution the carbonates and hydroxides can be discarded.

EQUATIONS FOR REACTIONS IN STEP 62. A sodium carbonate solution contains carbonate ions, which react with water to form hydroxide ions (see p. 28):

$$CO_3^{-2} + H_2O \rightleftharpoons HCO_3^- + OH^-.$$

Whether a metal ion will form a precipitate with CO_3^{-2} or OH^- in a sodium carbonate solution depends on the relative solubilities of the carbonate and the hydroxide. The one which will be less soluble in the sodium carbonate solution will precipitate. Calcium ion (Ca^{+2}) is certain to form the carbonate, because calcium carbonate is very much less soluble than calcium hydroxide:

$$Ca^{+2} + CO_3^{-2} \rightarrow \underline{CaCO_3} \text{ (White)}.$$

On the other hand, aluminum ion (Al^{+3}) will form a hydroxide precipitate:

$$Al^{+3} + 3OH^- \rightarrow \underline{Al(OH)_3} \text{ (White)}.$$

Aluminum hydroxide is very much less soluble than aluminum carbonate, which cannot exist in water solutions; instead aluminum hydroxide is formed in solutions where the pH is increased by carbonate ions (see equation above).

Whether zinc ion (Zn^{+2}) will form $Zn(OH)_2$ only, or $ZnCO_3$ only, or a mixture of both is dependent on the concentrations of carbonate and hydroxide ions and the solubility product constants of $Zn(OH)_2$ and $ZnCO_3$. In general, the divalent ions of metals can be considered as forming carbonates, like calcium, and the trivalent ions of metals can be considered as forming hydroxides, like aluminum (see problem 8, p. 157). A saturated sodium carbonate solution has a carbonate ion concentration of about 2 M, and the hydroxide concentration is about 0.02 M (see "hydrolysis," p. 28).

Typical reactions of salts of divalent ions with carbonate are:

$$Ca_3(PO_4)_2 + 3CO_3^{-2} \rightarrow \underline{3CaCO_3} + 2PO_4^{-3}.$$
$$ZnSO_4 + CO_3^{-2} \rightarrow \underline{ZnCO_3} + SO_4^{-2}.$$

Typical of the trivalent ions is the reaction of aluminum ion:

$$Al^{+3} + 3OH^- \rightarrow \underline{Al(OH)_3}.$$

If the carbonates and hydroxides of metal ions formed in this step are to be examined for the metal ions, they are dissolved in nitric acid. Hydrochloric acid may be used, but silver and lead carbonate are then converted to insoluble chlorides. Typical equations for reactions between hydronium ions of the nitric acid solution and the solid are:

$$CaCO_3 + 2H_3O^+ \rightarrow Ca^{+2} + CO_2\uparrow + 3H_2O.$$
$$ZnCO_3 + 2H_3O^+ \rightarrow Zn^{+2} + CO_2\uparrow + 3H_2O.$$
$$Al(OH)_3 + 3H_3O^+ \rightarrow Al^{+3} + 6H_2O.$$

Step 63. The Chloride Group Test. By making this test it is often possible to eliminate a large group of ions at once from those to be tested for in a sample. Silver nitrate solution is added to a solution of the sample. Then the following facts can be applied to help decide what ions may be present or absent: The salts of silver which are insoluble in water are silver borate, $AgBO_2$ (white); silver carbonate, Ag_2CO_3 (white); silver sulfite, Ag_2SO_3 (white); silver chloride, AgCl (white); silver bromide, AgBr (cream-colored); silver iodide, AgI (light yellow); silver phosphate, Ag_3PO_4 (yellow); silver arsenate, Ag_3AsO_4 (brown); silver chromate, Ag_2CrO_4 (red); and silver sulfide, Ag_2S (black). However, all but AgCl, AgBr, AgI, and Ag_2S dissolve in 3 M HNO_3 solution. Two different procedures are followed, Step 63A if the sample is soluble in water and Step 63B if it is not.

Step 63A. The Chloride Group Test on a Water-soluble Sample. To 5 drops of a solution of the sample in water only, like the solution from Step 59, add 1 drop of 0.5 M $AgNO_3$ solution. If no precipitate forms, all the anions listed in the paragraph above are absent and none of them need be tested for. The only ions that might be present are nitrate, NO_3^-, nitrite, NO_2^-, and acetate, $C_2H_3O_2^-$, all colorless. If a white precipitate forms, all colored ions are absent and need not be tested for. Only borate, carbonate, sulfite, and chloride will produce a white precipitate with silver ions. If the precipitate is black, sulfide is present and all the other ions listed above may also be present because the black of Ag_2S will hide any other colored or white precipitate that might be formed simultaneously.

Centrifuge the precipitate and discard the solution. To the solid add 5 drops of distilled water, stir, centrifuge, and discard the water. Now add 5 drops of 3 M HNO_3 solution to the solid and observe it closely. If it all dissolves, neither chloride, bromide, iodide, nor sulfide was present. If not all the precipitate dissolves, one or more of these four ions is present in the sample. The color of the solid will give an indication of what ion or ions for which to test. Centrifuge and decant the solution into a test tube. To the solution add 6 M ammonia solution a drop at a time until the solution tests alkaline to litmus; then add acetic acid until the solution tests acid to litmus and add 2 drops of 0.5 M $AgNO_3$ solution. If a white precipitate forms, either sulfite or borate may be present. If the precipitate is yellow, phosphate is probably present, and if it is brown or red, either arsenate or chromate is probably present and should be expected to give a positive test when tested for individually. *Caution:* a darker precipitate may darken the color of or hide a lighter-colored one. It is obvious that the absence of the above precipitate makes it unnecessary to make tests for any of those ions which would form precipitates here.

Step 63B. The Chloride Group Test on a Water-insoluble Sample. To 5 drops of the test solution prepared in Step 62 (p. 133), add 3 M HNO_3 until the solution tests acid; then add 2 drops of concentrated HNO_3 and 2 drops of 0.5 M $AgNO_3$ solution. If no precipitate forms, chloride, bromide, iodide, and sulfide are all absent and no further tests for these ions are needed. If the precipitate is black, all four of these ions must be tested for individually. If the precipitate is yellow, iodide is probably present, sulfide is absent, and iodide, bromide, and chloride must be tested for individually. If the precipitate is cream-colored, sulfide is absent and iodide may be absent, or present in a small amount, probably mixed with chloride. A precipitate, if at all yellowish, must be tested for chloride, bromide, and iodide. If the precipitate is pure white, chloride is present in the sample and no further test for any of these four ions is necessary.

Centrifuge and decant the solution into a test tube. Add concentrated ammonia to it drop by drop until the solution tests alkaline to litmus. If no precipitate forms, borate, phosphate, arsenate, and chromate are absent from the sample. Sulfite may have been present but if so, would have been oxidized to sulfate by nitric acid. An individual test for sulfite must be made according to the directions given in Step 72 (p. 145). Only borate gives a white precipitate at this point if sulfite is all oxidized. The colors of the other possible precipitates are given in Step 63A.

Equations for Reactions in Steps 63, 63A, and 63B. Only a few examples of equations for typical reactions need be given. Precipitation reactions:

$$Ag^+ + Br^- \rightarrow \underline{AgBr} \text{ (Cream-colored).}$$

$$3Ag^+ + PO_4^{-3} \rightarrow \underline{Ag_3PO_4} \text{ (Yellow).}$$

$$Ag^+ + BO_2^- \rightarrow \underline{AgBO_2} \text{ (White).}$$

$$2Ag^+ + CrO_4^{-2} \rightarrow \underline{Ag_2CrO_4} \text{ (Red).}$$

Some of the precipitates will not form from acid solution and if present will dissolve in acid. Examples:

$$Ag_3PO_4 + 2H_3O^+ \rightarrow 3Ag^+ + H_2PO_4^- + 2H_2O.$$
$$AgBO_2 + H_3O^+ \rightarrow Ag^+ + HBO_2 + H_2O.$$
$$2Ag_2CrO_4 + 2H_3O^+ \rightarrow 4Ag^+ + Cr_2O_7^{-2} + 3H_2O.$$

Oxidation of sulfite by nitric acid:

$$SO_3^{-2} + 2NO_3^- + 4H_3O^+ \rightarrow SO_4^{-2} + 2NO_2\uparrow + 6H_2O.$$

Step 64. The Sulfate Group Test. This test, like the chloride group test, may help eliminate several anions from among the ones to be tested for individually.

If the sample is soluble in water, take 5 drops of a solution of the sample in water from Step 59. If the sample is not soluble in water, take 5 drops of the solution prepared in Step 62 (p. 133) and add enough 3 M HNO_3 to make the solution acid to litmus. If carbonate in the solution forms carbon dioxide as the acid is added, boil out the CO_2. SO_2 from sulfite, if sulfite is present (note odor), will be boiled out also. Now add 3 M NH_3 solution a drop at a time to either of the two solutions until it is alkaline to litmus. Should a precipitate form, centrifuge and discard the solid, which will be hydroxides of cations. To the alkaline solution add 2 drops of 0.3 M barium chloride and 2 drops of 0.3 M calcium chloride solution. If no precipitate forms, sulfate (SO_4^{-2}), borate (BO_2^-), chromate (CrO_4^{-2}), phosphate (PO_4^{-3}), arsenate (AsO_4^{-3}), fluoride (F^-), and oxalate ($C_2O_4^{-2}$) are absent and need not be tested for further. If a water solution of the sample was used for this test and no precipitate formed, carbonate (CO_3^{-2}) and sulfite (SO_3^{-2}), in addition to those ions already listed, are absent and need not be tested for further. If the test solution from Step 62 was used, neither carbonate nor sulfite can be present to give a precipitate because both were eliminated when nitric acid was added and CO_2 and SO_2, if present, were boiled out.

If a yellow precipitate forms, chromate is probably present, and should later tests fail to indicate its presence in the sample, the later tests should be checked very carefully. Chromate is the only ion in the sulfate group which will give a colored precipitate.

If any precipitate forms, add 6 M HCl solution until the solution is acid; then add 5 drops more. If all the precipitate dissolves, sulfate is absent from the sample, but each of the other ions of the sulfate group must be tested for individually. If none of the precipitate dissolves in HCl solution, sulfate only, of the sulfate group ions, is present. However, it is very difficult, if not impossible, to be sure that *none* of a precipitate dissolves. It is safer to make individual tests for all ions of the group, even if the precipitate seems completely insoluble in acid.

EQUATIONS FOR REACTIONS IN STEP 64. The first group of reactions is precipitation by union of barium or calcium ions with each of the ions in the sulfate group:

$$Ba^{+2} + SO_4^{-2} \rightarrow \underline{BaSO_4} \text{ (White)}.$$
$$Ba^{+2} + 2BO_2^{-} \rightarrow \underline{Ba(BO_2)_2} \text{ (White)}.$$
$$Ba^{+2} + CrO_4^{-2} \rightarrow \underline{BaCrO_4} \text{ (Yellow)}.$$
$$3Ba^{+2} + 2PO_4^{-3} \rightarrow \underline{Ba_3(PO_4)_2} \text{ (White)}.$$
$$3Ba^{+2} + 2AsO_4^{-3} \rightarrow \underline{Ba_3(AsO_4)_2} \text{ (White)}.$$
$$Ca^{+2} + C_2O_4^{-2} \rightarrow \underline{CaC_2O_4} \text{ (White)}.$$
$$Ca^{+2} + 2F^{-} \rightarrow \underline{CaF_2} \text{ (White)}.$$

If the water solution of the sample was used for making the sulfate group test, carbonate and sulfite may also be present to react with either calcium or barium ions. With barium ions the equations are:

$$Ba^{+2} + SO_3^{-2} \rightarrow \underline{BaSO_3} \text{ (White)}.$$
$$Ba^{+2} + CO_3^{-2} \rightarrow \underline{BaCO_3} \text{ (White)}.$$

The second group of possible reactions is between hydronium ions, from the acid added, and the precipitate to form a relatively weak acid or, as in the case of chromate, an ion stable in acid solution:

$$Ba(BO_2)_2 + 2H_3O^{+} \rightarrow Ba^{+2} + 2HBO_2 + 2H_2O.$$
$$2BaCrO_4 + 2H_3O^{+} \rightarrow 2Ba^{+2} + Cr_2O_7^{-2} + 3H_2O.$$
$$Ba_3(PO_4)_2 + 4H_3O^{+} \rightarrow 3Ba^{+2} + 2H_2PO_4^{-} + 4H_2O.$$
$$Ba_3(AsO_4)_2 + 4H_3O^{+} \rightarrow 3Ba^{+2} + 2H_2AsO_4^{-} + 4H_2O.$$

These last two equations may also be written:

$$Ba_3(PO_4)_2 + 6H_3O^{+} \rightarrow 3Ba^{+2} + 2H_3PO_4 + 6H_2O.$$
$$Ba_3(AsO_4)_2 + 6H_3O^{+} \rightarrow 3Ba^{+2} + 2H_3AsO_4 + 6H_2O.$$

The acids H_3PO_4 and H_3AsO_4 are products as well as $H_2PO_4^{-}$ and $H_2AsO_4^{-}$.

The calcium salts react as follows:

$$CaF_2 + 2H_3O^{+} \rightarrow Ca^{+} + 2HF\uparrow + 2H_2O.$$
$$CaC_2O_4 + 2H_3O^{+} \rightarrow Ca^{+2} + H_2C_2O_4 + 2H_2O.$$

If the water solution of the sample was used for making the test, sulfite and carbonate will, if present, react as follows:

$$BaSO_3 + 2H_3O^+ \rightarrow Ba^{+2} + SO_2\uparrow + 3H_2O.$$
$$BaCO_3 + 2H_3O^+ \rightarrow Ba^{+2} + CO_2\uparrow + 3H_2O.$$

Tests for Individual Anions

The following steps are tests to be made either on the water solution of the sample or on the solution prepared as described in Step 62 (p. 133). In analyzing a sample the only tests to be made are for anions not eliminated by results of Steps 58, 59, 60, 61, 63, and 64.

Step 65. The Test for Borate. Borate as BO_2^- or $B_4O_7^{-2}$ occurs in many washing and cleaning preparations, usually as the sodium salt. In acid solution either ion is converted to some degree to orthoboric acid, H_3BO_3, which reacts with methyl alcohol to produce volatile methyl borate. The latter burns with a characteristic green flame.

Place a bit of the original sample, about one-fourth the size of a grain of rice, in a casserole, evaporating dish, or crucible. Add 2 drops of water; then, cooling after the addition of each drop, add 2 drops of concentrated sulfuric acid. Mix the acid with the sample and, under a hood, add 10 drops of methyl alcohol, CH_3OH, and stir again. Leaving the dish under the hood but in a good light, ignite the alcohol. If a green flame can be seen at once, the sample contained borate. However, if the green flame appears after about 30 seconds, it may be due to copper or barium.

If borate is found to be present in the sample and the sample is a salt or mixture of salts, borate must be eliminated before beginning Step 24, precipitation of Group III cations. Step 81 (p. 153) gives directions for eliminating borate before starting Step 24.

EQUATIONS FOR REACTIONS IN STEP 65. The first equations are for reactions between various possible borate ions and hydronium ions and water in the sulfuric acid solution, to produce orthoboric acid, H_3BO_3:

$$\text{(Metaborate) } BO_2^- + H_3O^+ \rightarrow H_3BO_3.$$
$$\text{(Tetraborate) } B_4O_7^{-2} + 2H_3O^+ + 3H_2O \rightarrow 4H_3BO_3.$$
$$\text{(Orthoborate) } BO_3^{-3} + 3H_3O^+ \rightarrow H_3BO_3 + 3H_2O.$$

The next reaction is between orthoboric acid and methyl alcohol, to form methyl borate, a gas:

$$H_3BO_3 + 3CH_3OH \rightarrow (CH_3O)_3B\uparrow + 3H_2O.$$

This reaction is catalyzed by acid to some degree.

The final reaction, the burning of methyl borate, probably produces several products, among them CO_2, H_2O, and some oxide or hydrated oxide of boron.

Step 66. The Test for Oxalate. This test is based on the fact that oxalate ions in an acid solution are oxidized to CO_2 by permanganate. The permanganate, which is an intense reddish purple, is reduced to colorless manganous ions.

To 2 drops of the anion test solution prepared in Step 62 (p. 133), add 1 drop of 6 M acetic acid and 1 drop of 0.3 M $CaCl_2$ (calcium chloride) solution. If a precipitate forms, it may be calcium oxalate or calcium fluoride, or both. If there is no precipitate, oxalate is absent. Centrifuge and wash the precipitate once with 5 drops of water. Centrifuge and discard the wash water. To the precipitate add 2 drops of water and 2 drops of 6 M H_2SO_4. Warm gently to boiling and cool until the test tube can barely be held in the hand. While it is still quite warm, add a drop of 0.1 M $KMnO_4$ (potassium permanganate) solution. If the color of the permanganate fades within 30 seconds, oxalate was present in the sample.

If the sample is a salt or a mixture of salts to be analyzed for both cations and anions, oxalate must be eliminated before starting Step 24 in the cation analysis. Directions for eliminating oxalate are given in Step 82 (p. 154).

EQUATIONS FOR REACTIONS IN STEP 66. The addition of calcium ions in calcium chloride solution causes calcium oxalate to precipitate:

$$Ca^{+2} + C_2O_4^{-2} \rightarrow \underline{CaC_2O_4} \text{ (White).}$$

Sulfuric acid converts calcium oxalate to oxalic acid:

$$CaC_2O_4 + H_2SO_4 \rightarrow \underline{CaSO_4} + H_2C_2O_4.$$

Permanganate oxidizes the oxalic acid to CO_2:

$$2MnO_4^- + 5H_2C_2O_4 + 6H_3O^+ \rightarrow 2Mn^{+2} + 10CO_2\uparrow + 14H_2O.$$

Some authors prefer an equation for the reaction between oxalate ions and permanganate:

$$2MnO_4^- + 5C_2O_4^{-2} + 16H_3O^+ \rightarrow 2Mn^{+2} + 10CO_2\uparrow + 24H_2O.$$

Step 67. The Test for Fluoride. Fluoride is converted to hydrofluoric acid, HF, which etches glass by reacting with silica, SiO_2, in glass to form SiF_4, a gas.

Clean a small watch glass thoroughly; it must be free of all grease and etch marks. Place a small bit of powdered sample on the watch glass in an area as small as possible. Place on top of the sample an amount of powdered ammonium sulfate, $(NH_4)_2SO_4$, equal in volume to that of the sample. Dip a small or pointed stirring rod into a solution made by mixing equal volumes of concentrated sulfuric acid and water. *Caution:* Add acid slowly, with stirring, to water — *not* water to acid — to avoid spattering. Add the small drop of acid on the end of the rod to the sample on the watch glass without disturbing it any more than necessary. Continue adding acid, drop by drop, to the pile, if necessary, until the pile of sample and ammonium sulfate is moistened, but not wet. Place the watch glass on a beaker with water in it and bring the water to a boil. Heat the watch glass about an hour, then set the watch glass and sample aside overnight. At a later day wash the sample away, dry the glass, and examine it closely, using a magnifying glass if possible. If the glass is etched where the sample contacted it, fluoride was present in the sample.

If the sample contained fluoride and it is a salt or mixture of salts, fluoride must be eliminated before beginning Step 24, precipitation of Group III cations. Step 82 (p. 154) gives directions for eliminating fluoride before starting Step 24.

EQUATIONS FOR REACTIONS IN STEP 67. The first reaction in this test is that between sulfuric acid and fluoride ions to produce hydrofluoric acid:

$$H_2SO_4 + F^- \rightarrow HSO_4^- + HF\uparrow .$$

The HF formed then reacts with silica in the glass:

$$4HF + SiO_2 \rightarrow SiF_4\uparrow + 2H_2O.$$

Because SiF_4 is a gas which will dissolve in the acid solution, silica is removed from the glass and the glass appears roughened or etched.

Step 68. The Tests for Arsenate and Phosphate. Both of these ions are tested for with the same reagent, magnesia mixture. If the results of Steps 63 and 64 do not indicate that phosphate and arsenate are definitely absent, proceed as follows.

To 3 drops of a solution made by dissolving 0.1 g of solid sample in 1 ml of water or to 3 drops of the solution from Step 62 (p. 133) if the sample is not soluble in water, add a drop of saturated sodium carbonate solution. Stir and test to see if the solution is alkaline to litmus. If it is not, add more sodium carbonate solution until an alkaline test is obtained. If a precipitate forms, centrifuge and discard the solid. The solid will be a carbonate or hydroxide of any metal ions except sodium, potassium, or ammonium. To the clear alkaline solution add concentrated HCl solution a small amount at a time until, on stirring, the solution tests acid; then boil out the CO_2. Add concentrated ammonia solution a very little at a time down the side of the tube, stirring and removing a little of the solution on the stirring rod for testing with litmus. When the test is just alkaline, add 3 drops of magnesia mixture and observe closely. If no precipitate forms in a few minutes, both arsenate and phosphate are absent. If a white precipitate forms, it is either $MgNH_4PO_4 \cdot 6H_2O$ or $MgNH_4AsO_4 \cdot 6H_2O$, or both, and the presence of either arsenate or phosphate or both is indicated, unless arsenic has been proved absent in Step 21 (p. 78). If arsenic was found absent in Step 21, then the precipitate indicates that phosphate is present in the sample, and no further test for phosphate or arsenate is needed.

If a precipitate forms and arsenic has not been proved absent in Step 21, centrifuge, discard the liquid, wash the precipitate three times with 5 drops of water each time, and add a drop of 0.5 M $AgNO_3$ solution to the precipitate from which the wash water has been removed. If arsenate is in the precipitate, it will be converted to a reddish-brown Ag_3AsO_4, and it will be necessary to test for phosphate by Step 69 because the color of Ag_3AsO_4 hides the light-yellow color of Ag_3PO_4, if it is present. If the color of the precipitate with $AgNO_3$ is yellow, arsenate is absent but phosphate is present and no further tests need to be made for either ion. If the precipitate is small and white after the $AgNO_3$ is added, both phosphate and arsenate are absent and need not be tested for further. A small amount of precipitate may be due to a trace of chloride.

If phosphate is found to be present in a salt or mixture of salts, the phosphate must be removed before beginning the precipitation of Group III cations in Step 24. Directions for doing this are given in Step 83 (p. 155). Although arsenate would also interfere in the same way as phosphate, it is removed in Step 7 or 7A, the Group II precipitation, where the arsenate is reduced and precipitated as As_2S_3 or else is converted to As_2S_5, depending on the conditions.

Equations for Reactions in Step 68. The addition of sodium carbonate provides carbonate and hydroxide ions to precipitate all cations but sodium, potassium, and ammonium. The equations for these reactions are given in Step 62 (p. 133).

Addition of HCl solution eliminates carbonate by forming CO_2 by the reaction between hydronium ions of the acid solution and carbonate ions:

$$2H_3O^+ + CO_3^{-2} \rightarrow CO_2\uparrow + H_2O.$$

The addition of ammonia solution adjusts the pH of the solution so that the precipitates expected to form if arsenate and phosphate are present will be insoluble. Ammonia solution, by reacting with hydronium ions of the acid, furnishes ammonium ions:

$$NH_3 + H_3O^+ \rightarrow NH_4^+ + H_2O.$$

The ammonium ions are an essential part of the precipitate.

The magnesium ions added in magnesia mixture react with ammonium and phosphate or arsenate to give white precipitates:

$$Mg^{+2} + NH_4^+ + PO_4^{-3} + 6H_2O \rightarrow \underline{MgNH_4PO_4 \cdot 6H_2O} \text{ (White).}$$
$$Mg^{+2} + NH_4^+ + AsO_4^{-3} + 6H_2O \rightarrow \underline{MgNH_4AsO_4 \cdot 6H_2O} \text{ (White).}$$

The addition of silver ions in silver nitrate to these white solids converts them to the silver salts:

$$MgNH_4PO_4 \cdot 6H_2O + 3Ag^+ \rightarrow Mg^{+2} + NH_4^+ + 6H_2O + \underline{Ag_3PO_4} \text{ (Yellow).}$$
$$MgNH_4AsO_4 \cdot 6H_2O + 3Ag^+ \rightarrow Mg^{+2} + NH_4^+ + 6H_2O + \underline{Ag_3AsO_4} \text{ (Red-brown or reddish).}$$

Step 69. The Test for Phosphate When Arsenate Is Present. If arsenate is found to be present by the procedure in Step 68, arsenic must be removed by precipitation of As_2S_5 by the same method as is used for precipitating As_2S_3 in Step 7 or 7A (p. 63). Then a test for phosphate can be made on the remaining solution.

Place a very small amount of original solid sample, about equal in volume to the head of a small pin, or 3 drops of a water solution of the sample, in a small test tube. Add 5 drops of concentrated HCl solution and heat the test tube in boiling water for 1 full minute. Add 3 drops of thioacetamide solution and again heat the tube in boiling water for 2 full minutes. Add about 5 drops of water, stir thoroughly, and centrifuge. If arsenate was present, a yellow precipitate of As_2S_5 should be in the bottom of the tube. Decant the liquid into a casserole or beaker and evaporate slowly to dryness, but do not heat after it is dry. Cool the dry material; add 4 drops of water and 2 drops of concentrated HNO_3. Stir to dissolve the dried salts. Centrifuge if not clear and decant the clear solution into another test tube. Add 2 drops of ammonium molybdate, $(NH_4)_2MoO_4$, solution and set the test tube in boiling water to heat for a full minute. Examine the solution closely. If a yellow precipitate can be seen, which often forms slowly on the walls of the test tube, it is $(NH_4)_3PO_4 \cdot 12MoO_3$, which indicates that phosphate was present in the sample. Arsenate would give a precipitate with the same appearance if it were not removed before making the test.

If phosphate is found to be present in either Step 68 or 69 and if the sample being analyzed is a salt or mixture of salts, phosphate must be removed from solution before Step 24, precipitation of Group III cations, is begun. Directions for doing this are given in Step 83 (p. 155).

EQUATIONS FOR REACTIONS IN STEP 69. The first reaction of importance is that of hydronium ions from hydrochloric acid solution with arsenate ion to produce some As^{+5} ions by the equilibrium reaction:

$$AsO_4^{-3} + 8H_3O^+ \rightleftharpoons As^{+5} + 12H_2O.$$

Then As^{+5} and sulfide ions from thioacetamide solution produce As_2S_5:

$$2As^{+5} + 5S^{-2} \rightarrow \underline{As_2S_5} \text{ (Yellow)}.$$

Some As_2S_3 (yellow) and sulfur may be formed also. The final reaction is between molybdate ions (MoO_4^{-2}), ammonium ions (NH_4^+), phosphate ions (PO_4^{-3}), and hydronium ions (H_3O^+) in the strongly acid solution.

$$3NH_4^+ + 12MoO_4^{-2} + PO_4^{-3} + 24H_3O^+ \rightarrow \underline{(NH_4)_3PO_4 \cdot 12MoO_3} + 36H_2O.$$

The ammonium molybdate solution is strongly alkaline with ammonia; yet the solution must be acid when the test is made. Therefore a large amount of nitric acid is required to neutralize ammonia in the ammonium molybdate solution and to make sure the concentration of hydronium ions is sufficiently great for the reaction to take place.

Tests for borate, fluoride, oxalate, and phosphate must be made before Step 24 (p. 86) is begun. The following tests can be made at any time, but are best made in the order listed here.

Step 70. The Test for Chromate. Omit this test if a solution of the sample in water is colorless or if chromium ions were not found to be present in the sample by the test for chromium in Step 38

(p. 100). Chromates are yellow in neutral solution and are changed to orange dichromate in strongly acid solutions. This test is essentially the same as is made for chromium in Step 38.

Place either 3 drops of a water solution of the sample or 3 drops of the test solution prepared in Step 62 (p. 133) in a test tube and add 1 M nitric acid a drop at a time, testing the solution with litmus after each drop of acid is added. As soon as the solution is acid, add 3 drops of 3 M HNO_3 and enough ether to make a layer of ether above the water in the test tube about $\frac{1}{4}$ to $\frac{1}{2}$ inch thick. Add 3 drops of 3% hydrogen peroxide (H_2O_2) solution, shake once quickly, and look for a blue color in the ether layer, which would indicate the presence of chromate.

Chromate in the sample will be reduced by sulfide ions in acid solution to Cr^{+3} and will appear in Group III just as if chromium had been in the sample as chromic ions (Cr^{+3}).

The equations for reactions in this test are given in Step 38 (p. 100).

Step 71. The Test for Sulfate. This test was made in the sulfate group test, Step 64 (p. 137), but it can be made on any solution at any time.

To 3 drops of a water solution of the sample or to 3 drops of the test solution prepared in Step 62 (p. 133), add 6 M HCl solution a drop at a time until the solution tests acid; then add 4 drops more. Add a drop of 0.2 M $BaCl_2$ (barium chloride) solution. A white precipitate indicates that sulfate was present in the sample. A very slight precipitate may be due to sulfate as an impurity, especially if sulfite, sulfide, or other sulfur compounds are in the sample. The solution from this test may be used for the sulfite test in Step 72.

Equation for the Reaction in Step 71. There is only one reaction of importance in this step, the combination of ions to form the $BaSO_4$ precipitate:

$$Ba^{+2} + SO_4^{-2} \rightarrow \underline{BaSO_4} \text{ (White)}.$$

Step 72. The Test for Sulfite. Sulfite, SO_3^{-2}, reacts with acids to produce sulfur dioxide, SO_2, a gas with a pungent odor easily recognized by anyone familiar with it. When sulfuric acid was added to some of the original sample in Step 61 (p. 130), if sulfite was present, bubbles of SO_2 could be seen forming rapidly where acid and sample came together. If there was no effervescence (bubbling) with the sulfuric acid treatment, the following test may be omitted.

In a test tube place 3 drops of a water solution of the sample or 3 drops of test solution prepared in Step 62 (p. 133) and add 6 M HCl solution drop by drop until the solution tests acid to litmus. Then add 4 drops more and also 3 drops of 0.2 M $BaCl_2$ solution. Centrifuge, decant the clear solution, and place it in another test tube. Add another drop of $BaCl_2$ solution. If more precipitate forms, centrifuge again and add another drop of $BaCl_2$ solution to the clear solution. Continue adding barium chloride and centrifuging until no more precipitate forms when $BaCl_2$ is added. All sulfate must be removed from the solution by precipitation as barium sulfate before the test for sulfite is made. To the clear solution, to which barium chloride solution was added without any precipitate forming, add a drop of 3% hydrogen peroxide (H_2O_2) solution. Warm the solution to near boiling. If a white precipitate forms, it is barium sulfate, the sulfate being formed by the oxidation of sulfite (SO_3^{-2}) to sulfate (SO_4^{-2}) by peroxide.

Equations for Reactions in Step 72. There are three reactions of importance although, if sulfate is present, two of them are the same reaction. All sulfate is first eliminated from solution by precipitation as $BaSO_4$:

$$Ba^{+2} + SO_4^{-2} \rightarrow \underline{BaSO_4} \text{ (White)}.$$

Then sulfite is oxidized to sulfate by hydrogen peroxide:

$$SO_3^{-2} + H_2O_2 \rightarrow SO_4^{-2} + H_2O.$$

Finally the sulfate produced is precipitated as barium sulfate:

$$Ba^{+2} + SO_4^{-2} \rightarrow \underline{BaSO_4} \text{ (White)}.$$

Step 73. The Test for Carbonate. This test may be omitted if no effervescence was seen in Step 61 (p. 130), treatment of the sample with sulfuric acid. There are two procedures by which carbonates are detected: one when sulfite is not present and one when sulfite has been found present by the test in Step 72.

If sulfite was found to be absent in Step 72, prepare some pieces of apparatus such as two loops of wire or two medicine droppers so that when one is dipped into barium hydroxide solution a little drop of the liquid will be visible to an observer. Have two test tubes available. To one, add a small amount of sodium sulfate from the shelf as a blank, or "control." To the other test tube add a small bit of the original sample. To each test tube add 3 drops of 3 M HCl, quickly dip the wire loops or the medicine droppers into the barium hydroxide, $Ba(OH)_2$, solution, and insert the drops of barium hydroxide solution into the tubes. Arrange the apparatus so that air, which contains CO_2, contacts the droplet of $Ba(OH)_2$ solution as little as possible. A short medicine dropper has a rubber bulb which can cover the top of a test tube if the glass part is only about an inch long and all of the glass part is lowered into the test tube (see Fig. 7). Both tubes must be treated exactly alike. Warm the bottoms of the tubes, but not to the boiling point. Observe the droplets of barium hydroxide inside the two tubes closely. If the droplet of $Ba(OH)_2$ solution exposed to the air inside the test tube containing the sample becomes more cloudy or white than the one in the control, carbonate is present in the sample. A small amount of carbonate is difficult to detect, because CO_2 in the air will cause a droplet of $Ba(OH)_2$ solution to become cloudy after a time.

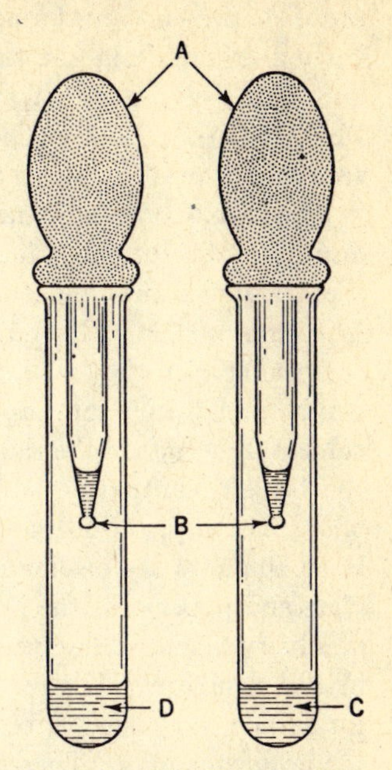

Fig. 7. The Test for Carbonate. A represents medicine droppers with short glass tubes and with a drop of $Ba(OH)_2$ solution (B) on the end of each. The bulbs of the droppers exclude air from the tubes while the test is in progress. C is the test material, D the control (Na_2SO_4).

If sulfite was found to be present in the sample in Step 72, add 3 drops of 3% hydrogen peroxide solution to both sample and control test tubes and stir both thoroughly; then add acid and complete the test as described above.

EQUATIONS FOR REACTIONS IN STEP 73. Carbonate reacts with hydronium ions from acid solutions to give carbon dioxide:

$$2H_3O^+ + CO_3^{-2} \rightarrow CO_2\uparrow + 3H_2O.$$

The CO_2 then reacts with barium hydroxide in solution to form insoluble barium carbonate:

$$Ba(OH)_2 + CO_2 \rightarrow H_2O + \underline{BaCO_3}\ \text{(White)}.$$

The equation for the oxidation of sulfite by peroxide is given in Step 72 (p. 146).

Step 74. The Test for Sulfide. Sulfides vary from highly soluble sodium sulfide to very insoluble sulfides such as HgS. Fortunately the very insoluble sulfides are usually colored or black, so that the preliminary inspection, solubility tests, etc., give an indication of the probable presence or absence of sulfide. The test depends on the conversion of sulfide in the sample to H_2S, a gas. The H_2S gas is allowed to come in contact with paper moistened with lead acetate, $Pb(C_2H_3O_2)_2$, solution. The dark-colored lead sulfide formed is easily seen and proves the presence of sulfide.

Place a bit of the original sample in a small beaker. Add 10 drops of 6 M HCl solution. Moisten a small piece of filter paper with lead acetate solution and place it on the convex side of a small watch glass. The moist paper will stick to the glass. Place the watch glass over the beaker with the moist paper inside the beaker. Warm the beaker gently and watch the paper. If it turns black or silvery black in color, sulfide was in the sample. As you remove the watch glass from the beaker, cautiously sniff the gases inside the beaker for the H_2S odor. If the paper did not change color, drop a few pieces of granulated zinc into the beaker and warm again as before, with the watch glass and paper over the beaker. If the paper changes color to black or silvery black, sulfide was in the sample in the form of some rather insoluble sulfide. If sulfite or sulfate is in the sample, a positive test for sulfide might occur, but only if the solution and zinc are heated too hot or too long. Therefore about 30 seconds, or a minute at most, of heating is all that is required. Because of this, some caution must be used in making this test if either sulfite or sulfate has been found present in the sample by the tests in Steps 71 and 72.

EQUATIONS FOR REACTIONS IN STEP 74. The more soluble sulfides, such as Na_2S and CaS, hydrolyze (see p. 129) with water to produce H_2S. Less soluble sulfides, such as FeS, react with acid solutions:

$$2H_3O^+ + FeS \rightarrow H_2S\uparrow + Fe^{+2} + 2H_2O.$$

Or hydronium ions react with sulfide ions directly:

$$2H_3O^+ + S^{-2} \rightarrow H_2S\uparrow + 2H_2O.$$

The more insoluble sulfides, such as NiS, HgS, and Ag_2S, do not react with hydronium ions in a nonoxidizing acid solution but do react with a strongly reducing combination of zinc and acid:

$$HgS + 2H_3O^+ + Zn \rightarrow Zn^{+2} + Hg + 2H_2O + H_2S\uparrow.$$

The final reaction of the test causes the change in color on the paper. Hydrogen sulfide reacts with lead acetate to produce lead sulfide, which is black:

$$H_2S + Pb(C_2H_3O_2)_2 \rightarrow 2HC_2H_3O_2 + \underline{PbS} \text{ (Black)}.$$

Step 75. The Test for Iodide. Iodide is easily oxidized in an acid solution to free iodine by nitrite ions, hydrogen peroxide, ferric nitrate, hypochlorite, or chlorine dissolved in water. The liberated iodine is brown when dissolved in water, but dissolved in carbon tetrachloride, CCl_4, it is violet. The violet color is so intense that a very dilute solution can be detected.

Place 6 drops of test solution as prepared in Step 62 (p. 133) in a test tube and add 6 M acetic acid a drop at a time until the solution tests acid to litmus. Then add 4 drops of CCl_4 and 2 drops of 0.2 M potassium nitrite, KNO_2, solution. Shake thoroughly and observe the color of the CCl_4 (lower) layer. If it is purplish in color, iodide was present in the solution. If no iodide is found to be present, the solution is ready for the test for bromide in Step 76.

If iodide is present, remove the CCl_4 solution of iodine from the bottom of the test tube with a long medicine dropper and discard it. Add 4 drops of fresh CCl_4 and a drop of 0.2 M KNO_2. Shake again. If a violet color is produced in the CCl_4 layer again, add fresh CCl_4 and KNO_2 until no violet color of iodine appears in the CCl_4 layer. The iodide is now all oxidized to iodine and removed from the water layer by extraction. The water layer is now ready for Step 76, the test for bromide.

EQUATION FOR THE REACTION IN STEP 75. Nitrite oxidizes iodide to free iodine in a slightly acid solution.

$$2NO_2^- + 4H_3O^+ + 2I^- \rightarrow 2NO\uparrow + I_2 + 6H_2O.$$

The iodine produced dissolves in CCl_4 without reacting.

Step 76. The Test for Bromide. The bromide test is like the iodide test except that a stronger oxidizing agent is required. The free bromine dissolves in a layer of CCl_4 to give a brown or yellow solution.

Place the solution from Step 75, or if it is not available, some of the test solution from Step 62 (p. 133) in a test tube. Add an equal volume of concentrated (16 M) HNO_3. Dip the test tube in boiling water for 30 seconds and cool in cold water. Add 2 drops of CCl_4 and shake thoroughly. If the CCl_4 layer becomes yellowish or reddish brown, bromide was present in the sample. Discard the CCl_4 layer if the test is positive and add 5 drops of CCl_4, shake thoroughly, and again discard the CCl_4 layer. The solution that remains should be free of bromine and is now ready for Step 77, the test for chloride. If the test for bromide indicated that bromide was absent from the sample, the water layer is ready for Step 77 without further treatment.

Equation for the Reaction in Step 76. The strong nitric acid oxidizes bromide, but not chloride, if the solution is not heated too long:

$$2NO_3^- + 4H_3O^+ + 2Br^- \rightarrow 2NO_2\uparrow + Br_2\uparrow + 6H_2O.$$

Bromine dissolves in CCl_4 without any reaction.

Step 77. The Test for Chloride. This test applies the familiar precipitation of silver chloride, AgCl.

To the solution from Step 76, or to 3 drops of the test solution from Step 62 (p. 133), made acid with 6 M HNO_3, add 4 drops more of the HNO_3, 10 to 15 drops of water, and 2 drops of 0.5 M silver nitrate solution. A white precipitate (silver chloride) indicates the presence of chloride in the sample. Compare the color of the precipitate with a piece of white paper. If all the bromide or iodide was not removed, the precipitate will be yellowish. It will then be necessary to repeat Steps 75 and 76 with a fresh sample, using care to be certain to remove all the iodide and bromide before the chloride test is made.

Equation for the Reaction in Step 77. The only reaction is the formation of silver chloride from its ions:

$$Ag^+ + Cl^- \rightarrow \underline{AgCl} \text{ (White).}$$

Step 78. The Test for Acetate. This test depends on the formation of volatile ethyl acetate, a compound which has a characteristic sweetish odor. Tests which depend on odors are often unreliable. A simultaneous test should be made on a small sample of sodium acetate from the shelf in order to be able to compare the odor from your sample with that from a sample known to contain acetate.

Place a small amount of original sample in a crucible, evaporating dish, or small beaker. Add 2 drops of 1–1 sulfuric acid (one part water, one part concentrated acid, the acid added to the water). If acetate is present in large amount, the familiar odor of vinegar can be detected at this point and no further proof of the presence of acetate is needed. If no odor of vinegar can be detected, add 2 drops of ethyl (*not* methyl) alcohol and stir well. Warm gently; when hot but not boiling, remove from heat and sniff cautiously. **Do not** smell the mixture while it is being heated because it contains sulfuric acid and might spatter. If the same sweetish odor of ethyl acetate can be detected coming from the sample as from the control, known to contain acetate, report that the sample contained acetate.

EQUATIONS FOR REACTIONS IN STEP 78. The first reaction is between sulfuric acid and acetate to produce acetic acid:

$$C_2H_3O_2^- + H_2SO_4 \rightarrow HC_2H_3O_2 + HSO_4^-.$$

Acetic acid then reacts, in the presence of sulfuric acid, with ethyl alcohol to produce ethyl acetate:

$$HC_2H_3O_2 + C_2H_5OH \rightarrow C_2H_5O_2C_2H_3O_2 + H_2O.$$

The latter evaporates when warmed and is detected by its odor.

Step 79. The Test for Nitrite. Evidence that this ion is present will be found when sulfuric acid is added to the sample in Step 61 (p. 130), the preliminary test with H_2SO_4. Cold sulfuric acid produces nitrogen dioxide, NO_2, a brown gas, with nitrite:

$$2NO_2^- + H_2SO_4 + 2H_3O^+ \rightarrow 2NO_2\uparrow + SO_2\uparrow + 4H_2O.$$

If no trace of brown gas is observed in Step 61, this test may be omitted, because nitrite is not present. This test depends on the ability of nitrite to oxidize iodide to free iodine in a slightly acid solution.

Place 3 drops of test solution prepared as in Step 62 (p. 133) in a test tube. Add 3 drops of water, 3 drops of 6 M acetic acid, 3 drops of 2 M potassium or sodium acetate, 1 drop of 0.5 M potassium iodide solution, and 4 or 5 drops of CCl_4. Shake for 10 seconds and if the CCl_4 layer becomes violet in color as a result of dissolving free iodine, nitrite was present in the sample. Nitrate will not interfere by giving the same result.

EQUATION FOR THE REACTION IN STEP 79. The reaction is oxidation of iodide by nitrite, the same reaction as was employed to test for iodide in Step 75 (p. 149):

$$2NO_2^- + 2I^- + 4H_3O^+ \rightarrow 2NO\uparrow + I_2 + 6H_2O.$$

Step 80. The Test for Nitrate. This test depends on the formation of a brown unstable compound of ferrous ions combined with nitric oxide, $FeNO^{+2}$. If iodide, bromide, or chromate is present when the test is made, it produces a color which interferes by masking the brown color on which the test depends. Since all nitrates are somewhat soluble in water, whether the whole sample is soluble or not, a solution of that part of the sample which will dissolve in water is tested for nitrate after the chromate, iodide, or bromide is removed.

Place a small amount of the original sample, about equal in volume to a grain of rice, in a test tube. Add 10 drops of water and heat to boiling. If the solution is not clear, centrifuge and place the clear liquid in another test tube. If chromate, iodide, and bromide have been found to be absent, proceed with the test on this solution as directed in the next paragraph. If chromate was found to be present in Step 70, add to the clear solution 0.5 M $Pb(C_2H_3O_2)_2$ (lead acetate), drop by drop until further addition gives no more precipitate. Centrifuge and place the clear solution in another test tube. If iodide and bromide were found to be absent in Steps 75 and 76, this solution is ready for the test as directed in the next paragraph. If iodide or bromide or both were found to be present, add saturated silver sulfate drop by drop until no further precipitation occurs when more silver salt is added. Centrifuge and remove the clear solution to a clean test tube for testing as described in the next paragraph.

To the clear solution prepared as described in the paragraph above add enough 3 M sulfuric acid, with stirring, to make the solution just acid to litmus. Then allow about 10 drops of concentrated H_2SO_4 to run down the inside of the tube with as little agitation as possible so that the acid settles to the bottom, forming a layer. Set the tube in cold water, preferably ice water, to cool. While the test solution is cooling, place 10 drops of water, a few grains of ferrous sulfate, $FeSO_4 \cdot 7H_2O$, and 1 ml of concentrated H_2SO_4 in a test tube, stir, and cool for a few seconds. Add 5 or 6 drops of this freshly prepared ferrous sulfate solution to the cooled test solution, allowing it to run down the inside wall of the tube a drop at a time. This solution should form a layer directly over the H_2SO_4 layer. Set the tube in a stand for 15 to 20 minutes and occasionally observe the boundary between the sulfuric acid layer and the ferrous sulfate layer. A distinct brown line or ring at the boundary indicates the presence of nitrate in the sample. Make this test on a small amount of some nitrate sample obtained from the side shelf in order to become familiar with the appearance of this, the "brown ring" test.

EQUATIONS FOR REACTIONS IN STEP 80. Nitrate is reduced to NO by ferrous sulfate in the strongly acid solution:

$$NO_3^- + 3Fe^{+2} + 4H_3O^+ \rightarrow 3Fe^{+3} + 6H_2O + NO\uparrow .$$

The NO, soluble in cold water to some degree, combines with ferrous ions to form a brown complex ion:

$$Fe^{+2} + NO \rightarrow FeNO^{+2} \text{ (Brown)}.$$

A common error of students is to make this test on a sample to which nitric acid has been added. In such cases the test is usually positive.

The Elimination of Interfering Anions

The following three steps are directions for eliminating anions that cause errors in the analysis of Groups III and IV cations. The four anions borate, fluoride, oxalate, and phosphate form salts with practically all cations except Na^+, K^+, and NH_4^+, which are soluble in acids but insoluble in neutral or alkaline solutions. Therefore these anions cause no trouble in precipitating Group I or Group II cations where the solutions are acid. However, when Group III cations are precipitated the solution is made alkaline, and if Group IV cations are present and if borate, fluoride, oxalate, or phosphate is present at the same time, precipitation of calcium, strontium, barium, or magnesium salts of these anions will take place along with precipitation of sulfides and hydroxides of Group III cations. This causes separations and tests in the Group III analysis to be messy and uncertain, and also results in there being little, if any, of the Group IV cations left in solution to be precipitated in Step 41 (p. 105).

The following three steps should be done only if necessary and in the order given. It is unlikely that all four anions to be eliminated will be in one sample, but in case they are, do Step 81, then 82, and then 83. If only one of the four anions must be eliminated, do only the step which applies to that anion. If two anions must be eliminated, do only the two steps which apply in the order in which the steps are given. For example, if fluoride and phosphate must be eliminated, do only Steps 82 and 83; if borate and phosphate are the two, do only Steps 81 and 83.

Step 81. The Elimination of Borate as an Interfering Anion. If borate is found to be present in the sample in Step 65, it must be eliminated from the solution left from Step 7 or 7A, from which Group III cations are to be precipitated in Step 24 (p. 86).

Heat the solution from Step 7 or 7A to boiling in a small beaker and boil gently to remove all H_2S. As the volume of the solution decreases, add equal volumes of concentrated HCl and methyl alcohol to replenish the volume lost by evaporation until 5 full minutes of boiling time has elapsed. Then continue evaporating, adding a drop or two of methyl alcohol occasionally until the volume of liquid is no larger than about 2 drops. Continue the evaporation slowly without adding anything, until the residue is dry. Do not heat the residue after it is dry.

If oxalate or fluoride is to be eliminated, this residue is ready for Step 82. If both oxalate and fluoride are absent, dissolve the residue from evaporation in 1 ml of 0.3 M HCl solution. This solution is ready for Step 24 (p. 86) unless phosphate must be removed, in which case it is ready for Step 83.

The equations for reactions in this step are the same as those for the borate ion test, Step 65 (p. 139).

Step 82. The Elimination of Oxalate and Fluoride as Interfering Anions. This procedure eliminates both oxalate and fluoride at the same time. Boiling with a mixture of concentrated nitric and hydrochloric acids not only oxidizes oxalate to CO_2 but also produces HF from fluoride, which boils away during the evaporation.

If borate is absent from the sample, boil the solution from Step 7 or 7A in a Pyrex beaker until barely dry.

To the dried residue from either Step 7 or 7A or from Step 81, in a Pyrex beaker (*not* in a porcelain casserole), add 15 drops of concentrated HCl and 8 drops of concentrated HNO_3. Evaporate slowly under a hood until barely dry. Add 5 drops of concentrated HCl and 3 drops of concentrated HNO_3 and evaporate to dryness again. Do not heat the residue after it is just dry. Dissolve the residue in 15 drops of 0.3 M HCl solution by stirring the residue and acid solution together for a few minutes. If the residue is not dissolved after 4 or 5 minutes, add a drop of concentrated HCl and heat to boiling. If this treatment is not sufficient to dissolve all of the residue, add more HCl and heat again. Dilute the solution remaining to about 1 ml and the solution is ready for Step 24 (p. 86) unless phosphate is present in the sample, in which case the solution should be treated as described in Step 83.

Equations for Reactions in Step 82. Nitric acid oxidizes both oxalate and chloride. The reaction with oxalate is essential in eliminating oxalate. The reaction with chloride is incidental:

$$8H_3O^+ + 2NO_3^- + 3C_2O_4^{-2} \rightarrow 6CO_2 + 2NO\uparrow + 12H_2O.$$

$$8H_3O^+ + 2NO_3^- + 6Cl^- \rightarrow 3Cl_2 + 2NO\uparrow + 12H_2O.$$

Hydronium ions in the strongly acid solution react with fluoride ions to produce HF, hydrofluoric acid, which is a poisonous gas that evaporates along with water:

$$H_3O^+ + F^- \rightarrow H_2O \uparrow + HF \uparrow .$$

Step 83. The Elimination of Phosphate as an Interfering Anion. The separation of phosphate depends on the insolubility of $FePO_4$ in a solution in which the pH is controlled by adjusting the concentration of acetic acid and acetate ion with respect to each other. The acetic acid–acetate solution is therefore a buffer (see p. 36). Cr^{+3} and Al^{+3} are also precipitated, along with Fe^{+3}, as phosphates, if they are present and if there is sufficient phosphate in the sample to combine with them. Iron is tested for first; then, to be sure that there is sufficient ferric ion in solution to precipitate all of the phosphate, some more ferric ions are added whether the test for it was positive or not. The excess of Fe^{+3}, after separation of the $FePO_4$, a solid, is caused to precipitate out merely by boiling the solution. At the temperature of boiling water the hydroxide ion concentration is slightly higher than at room temperature, and basic ferric acetate, $Fe(OH)_2(C_2H_3O_2)$, forms as a solid and is removed by centrifuging.

For this procedure, if borate has been removed, use the solution from Step 81; if oxalate or fluoride was removed, use the solution from Step 82; if neither borate, oxalate, nor fluoride was removed, boil the solution from Step 7 or 7A until all H_2S is removed and use this solution.

To the appropriate solution add 3 drops of bromine water. Bromine oxidizes Fe^{+2} to Fe^{+3} and H_2S to free sulfur, if any remains after boiling. Boil the solution under a hood for a few minutes to remove most of the bromine, and test this solution for the presence of ferric iron before adding ferric ions to precipitate phosphate. To do this, remove 2 drops of the solution just treated with bromine and boiled, place it in a test tube, add 3 or 4 drops of water and 1 or 2 drops of 0.5 M KSCN (potassium thiocyanate) solution. If a red color develops, iron is in the test solution and must be reported as being in the sample being analyzed. The red iron compound is $Fe(SCN)^{+2}$. A very faint pink, or no change in color, indicates that iron was absent from the sample.

Whether iron was found to be present or not, place the solution remaining, after that for the iron test was removed, in a small beaker or casserole and add 6 M NH_3 solution a drop at a time until the solution is alkaline to litmus. Then add 3 drops more of the ammonia solution. To this alkaline solution add 6 M acetic acid a drop at a time until the solution is acid to litmus; then add 3 drops more of the acetic acid. If Fe^{+3}, Al^{+3}, or Cr^{+3} is in solution, a precipitate will form consisting of the phosphates of any or all of these ions. If the solution above the precipitate is deep red or red-brown, there was sufficient Fe^{+3} in the sample to precipitate all the phosphate in the sample. If the solution is not dark reddish, add 0.1 M $FeCl_3$ (ferric chloride) solution a drop at a time until a dark-reddish or red-brown solution is produced. The addition of Fe^{+3}, as ferric chloride, provides sufficient ferric ions to precipitate all phosphate with a small excess of ferric ions.

The excess of Fe^{+3} must be precipitated out, along with the precipitated phosphates. Boil the reddish-colored solution slowly until only about 0.5 ml remains; then add 2 ml of water, stir, and centrifuge. The solution may contain Ni^{+2}, Co^{+2}, Mn^{+2}, or Zn^{+2}. Thus Steps 30 and 31 in the nickel division can be omitted, and in the aluminum division only the test for zinc need be made on this solution.

The solid remaining may contain phosphates and basic acetates of Fe^{+3}, Cr^{+3}, and Al^{+3}. Transfer the solid to a casserole or beaker with about 1 ml of water, add 1 ml of 6 M NaOH solution and 3 drops of 3% hydrogen peroxide solution, and heat to boiling, but do not boil. After the solution has remained at near boiling for at least a minute, transfer to a test tube, and centrifuge. The solution may contain aluminate, AlO_2^-, and chromate, CrO_4^{-2}, and is ready for Step 35 (p. 97), except that Zn^{+2} cannot be present and Step 37 may therefore be omitted. The solid is $Fe(OH)_3$ and may be discarded.

EQUATIONS FOR REACTIONS IN STEP 83. Oxidation of ferrous ions to ferric ions by bromine is the first reaction of importance:

$$2Fe^{+2} + Br_2 \rightarrow 2Fe^{+3} + 2Br^-.$$

H_2S, if present, is oxidized by bromine, producing free sulfur and a water solution of HBr which becomes hydronium ions and bromide ions:

$$H_2S + Br_2 + 2H_2O \rightarrow 2H_3O^+ + 2Br^- + \underline{S} \text{ (White)}.$$

The reaction which is the test for ferric ions is the formation of the red, soluble complex ion $FeSCN^{+2}$ (see p. 94):

$$Fe^{+3} + SCN^- \rightarrow FeSCN^{+2}.$$

The red complex is ferric thiocyanate.

The precipitation of phosphate by formation of insoluble phosphates of trivalent ions is the principal reaction of the procedure. The equation for the reaction between ferric ions and phosphate is given as an example (but it must be remembered that this reaction takes place only if the pH is properly adjusted):

$$Fe^{+3} + PO_4^{-3} \rightarrow \underline{FePO_4}.$$

The precipitation of excess ferric ions by boiling the solution takes place because in the reaction:

$$H_2O + H_2O \rightleftharpoons H_3O^+ + OH^-$$

the equilibrium shifts toward the right at higher temperatures, increasing both the H_3O^+ and OH^- ion concentrations. The higher hydroxide ion concentration produces, in the presence of acetate ions, a combination with ferric ions that is insoluble. As a result, basic ferric acetate precipitates:

$$Fe^{+3} + 2OH^- + C_2H_3O_2^- \rightarrow \underline{Fe(OH)_2(C_2H_3O_2)} \text{ (Reddish brown).}$$

Review Questions and Problems

1. In analyzing a salt, why must the tests for ammonium, sodium, and potassium be made on the original sample?
2. What four anions must, if present, be eliminated before precipitating the Group III cations? Why?
3. What solid salts might effervesce with sulfuric acid?
4. What solid salts might produce the odor of H_2S with concentrated sulfuric acid?
5. If antimony nitrate is placed in water, why does a solid form if all nitrates are soluble in water? How can the solid be dissolved?
6. Why does ZnS react with HCl to produce H_2S whereas HgS does not?
7. Write equations for reactions of $Zn_3(PO_4)_2$, $Al(NO_3)_3$, and CaF_2 with sodium carbonate in solution.
8. If a zinc salt such as $ZnSO_4$ is boiled with a saturated sodium carbonate solution, which Zn compound would precipitate in greater quantity, $Zn(OH)_2$ or $ZnCO_3$? Saturated sodium carbonate solution is about 2 M in carbonate, and as a result of hydrolysis the hydroxide ion concentration is about 0.02 M.
9. For practice in learning how to analyze salts, choose any one cation and any one anion, assume they are a salt to be analyzed, and make a list of what would be seen by a student who did Steps 58 through 64.
10. Do the same thing as in question 10 but choose several cations and anions. In addition to the results of Steps 58 to 64, record the reactions

for each step needed to prove the presence of each cation or anion in a complete analysis.

11. A single solution contains Cl^-, PO_4^{-3}, S^{-2}, and CN^-, all 0.05 molar.
    a. What concentration of silver ions is required to saturate the silver salt of each ion?
    b. In what order will the ions precipitate if silver ions are added very slowly to the solution until all are precipitated?

10

Mathematical Operations

Significant Figures. Significant figures are those that have meaning and some degree of dependability. The number of significant figures in a number is found by counting from left to right, omitting zeros to the left of all digits but including all zeros to the right of all digits if the zeros to the right indicate the degree of reliability of the number and not just the position of the decimal point. The last significant figure may be one that is somewhat doubtful.

Examples. The following numbers have 3 significant figures: 217, 21.7, 2.17, 0.0217, and 0.000217.

The following numbers would generally be considered to have two significant figures for certain, but the second may have 3 and the third 4 significant figures; 43, 430, 4300. In 430 and 4300 the zeros are ambiguous because they may indicate the position of the decimal point or they may indicate the accuracy of the number and therefore the number of significant figures in the numbers. For the number 4300 to be written so as to indicate that it contains only 2 significant figures one can write 43×10^2 or 4.3×10^3. If it is written thus, the reader knows that one can rely on the number to be between 4200 and 4400, the most probable value being 4300.

To indicate that the value is accurately known to 2 figures but that the third figure is somewhat doubtful one can write 4.30×10^3. If it is written thus, the reader is confident that the true value of the number lies between 4290 and 4310, the most probable value being 4300.

Large and Small Numbers. In the example above, the number 4.30 was multiplied by 10^3, or 1000. The use of exponents of 10 to indicate the decimal point in writing large or small numbers can be extended so as to apply both positive and negative exponents of 10. The following list gives several numbers and their equivalents as powers of 10.

$$10{,}000 = 10^4.$$
$$1{,}000 = 10^3.$$
$$100 = 10^2.$$
$$10 = 10^1.$$
$$1 = 10^0.$$
$$0.1 = 10^{-1}.$$
$$0.01 = 10^{-2}.$$
$$0.001 = 10^{-3}.$$
$$0.0001 = 10^{-4}.$$

Examples. Some numbers and their equivalents using exponents of 10 are:

$$12{,}000 = 1.2 \times 10^4.$$
$$43{,}770{,}000 = 4.377 \times 10^7.$$
$$0.0031 = 3.1 \times 10^{-3}.$$
$$0.0000000642 = 6.42 \times 10^{-8}.$$

The number of significant figures in a number is easily expressed by writing numbers employing exponents of 10 to locate the decimal point. The number 12,000 as written above has 2 significant figures. If it were known to be between 11,990 and 12,010, this fact can be expressed by writing $12{,}000 = 1.200 \times 10^4$. When it is written thus, the reader knows the value to be accurate to 3 significant figures and that the fourth (the second zero) is probably correct but may be doubtful.

Logarithms. A *logarithm* is a number expressed as a power of a base number. There are two base numbers frequently used. The base for *common*, or *Napierian*, *logarithms* is 10 and the base for *natural logarithms* is e (e = 2.17828). A table for common logarithms of numbers is given in Appendix VIII, p. 176. Natural logarithms have special applications not required for solving problems in this book. The simplest logarithms are merely exponents of 10 ("log" is the abbreviation of logarithm):

$$100 = 10^2, \log 100 = 2.$$
$$10{,}000 = 10^4, \log 10{,}000 = 4.$$
$$0.001 = 10^{-3}, \log 0.001 = -3.$$
$$0.0000001 = 10^{-7}, \log 0.0000001 = -7.$$
$$1 = 10^0, \log 1 = 0.$$

To express numbers between 0 and 10, an exponent of 10 between 0 and 1 is required. This exponent of 10, the logarithm of the number, can be found in a table of logarithms (p. 176), where logarithms of numbers between 1 and 10 only, called *mantissas*, are listed.

Example 1. Find the logarithm of 3 in the table of logarithms. Look down the left-hand column of the logarithm table (labeled "N") to 30 and across to the column labeled "0." The value found is 0.4771, and it is the logarithm of 3.00. 10 taken to the 0.4771 power is equal to 3 ($10^{0.4771} = 3$).

Example 2. Find the logarithm of 3.43 in the table. Look down the "N" column to 34 and across to the column labeled "3" and 0.5353 is found. This is the logarithm of 3.43 ($10^{0.5353} = 3.43$).

Example 3. Find the logarithm of 3.72943. Look down column "N" to "37," across to column "2." The mantissa found is 0.5705, the logarithm for 3.72. In the column labeled "3," the logarithm for 3.73 is found to be 0.5717. The number 3.72943 rounded off to 3 places is 3.73. If only 3 significant figures are to be expressed, 0.5717 is taken as the logarithm of 3.72943. If 4 significant figures are to be indicated, the logarithm of 3.729 is desired. Because 3.729 is $\frac{9}{10}$ of the way between 3.72 and 3.73, the logarithm for it will be $\frac{9}{10}$ of the way from log 3.72 to log 3.73. The logarithm of 3.729 is thus found to be 0.5716. The process of calculating values between those given in the tables is called *interpolation.*

The logarithm of a number not between 1 and 10 must include a *characteristic* along with the mantissa. The characteristic is nothing more than the power of 10 required to give the proper decimal point to the number expressed by the mantissa alone. Negative characteristics are written with a line over them. If both the characteristic and the mantissa are negative, a negative sign precedes the characteristic. Some examples illustrate this.

$$\log 25 = \log(10^1 \times 2.5) = 1.3979.$$
$$\log 825 = \log(10^2 \times 8.25) = 2.9165.$$
$$\log 6.2 \times 10^{23} = \log(10^{23} \times 6.2) = 23.7924.$$
$$\log 0.0047 = \log(10^{-3} \times 4.7) = \bar{3}.6721 = -2.3279.$$
$$\log 3.674 \times 10^{-6} = \bar{6}.5651 = -5.4349.$$
$$\log 1.292 \times 10^{-3} = \bar{3}.1113 = -2.8887.$$

Calculating with Logarithms. Calculations are greatly simplified by applying logarithms in multiplication, division, and finding powers and roots.

MULTIPLICATION. To multiply numbers, add the logarithms of the numbers and find the antilogarithm of the sum. The *antilogarithm* is the number represented by the logarithm.

Example. Calculate the product of 274 multiplied by 0.8829.

Add logarithms:	log 274 =	2.4378.
	log 0.8829 =	$\bar{1}.9460.$
	Total	2.3838

The total, 2.3838, is the logarithm of the product sought. To find the antilogarithm, search for 3838, the mantissa, among those in the table until the mantissa nearest 3838 is found. 3838 is found in the column labeled "2," across from "24" in the column labeled "N." The number sought is then 2.42, the antilogarithm of 0.3838. The characteristic 2, in 2.3838, indicates that the product is 100 (10^2) times as large as 2.42. The product is therefore 242, as close to the true value, 241.9146, as a 4-place log table gives. A 6-place log table gives 241.91 as the answer.

DIVISION. To divide one number by a second, using logarithms, subtract the logarithm of the second from the logarithm of the first and find the antilogarithm of the result.

Example. Divide 4736 by 972.

$$\begin{aligned} \log 4736 &= 3.6754. \\ \log 972 &= \underline{2.9877.} \\ \text{Difference} \quad & \ 0.6877 \end{aligned}$$

The antilog of 0.6877 is 4.872.

COMBINED MULTIPLICATION AND DIVISION. Complicated combinations of multiplication and division become simple problems of addition and subtraction of logarithms of the numbers.

Example 1. Using logarithms calculate the value of the following fraction:

$$\frac{0.00472 \times 267 \times 66.4}{14{,}200 \times 0.369 \times 1.135}$$

Add the logarithms of all the factors:

$$\begin{aligned} \log 0.00472 &= \bar{3}.6739 \\ \log 267 &= 2.4265 \\ \log 66.4 &= \underline{1.8222} \\ \text{Totals} \quad & \ 1.9226 \end{aligned} \qquad \begin{aligned} \log 14{,}200 &= 4.1523 \\ \log 0.369 &= \bar{1}.5670 \\ \log 1.135 &= \underline{0.0550} \\ & \ 3.7743 \end{aligned}$$

Subtract the logarithms of the product below the line from the logarithm of the product above the line.

$$1.9226 - 3.7743 = \bar{2}.1483.$$

The characteristics and mantissas are subtracted as separate items. If the mantissa subtracted is larger than the one it is subtracted from, the characteristic of the answer is one unit more negative. The antilogarithm of $\bar{2}.1483$ is $1.41 \times 10^{-2} = 0.0141$. If the answer is one where 4 significant figures are required, an interpolation between 0.1461 and 0.1492 can be made. The mantissa for the answer is 0.1483, which is $\frac{22}{31}$, or about 0.7, of the way from 0.1461 to 0.1492. The antilogarithm can be estimated to be about 0.7 of the way from 1.40 to 1.41, or 1.407.

Example 2. Solve using logarithms:

$$\frac{464.27 \times 0.00207 \times 0.0218}{2.48 \times 3.86}$$

$$\begin{aligned} \log 464.27 &= 2.6668. \\ \log 0.00207 &= \bar{3}.3160. \\ \log 0.0218 &= \bar{2}.3385. \\ \text{Totals} &\quad \bar{2}.3213 \end{aligned} \qquad \begin{aligned} \log 2.48 &= 0.3945. \\ \log 3.86 &= 0.5866. \\ & \quad 0.9811 \end{aligned}$$

Subtracting logarithms to divide:

$$\bar{2}.3213 - 0.9811 = \bar{3}.3402.$$

The antilogarithm of $\bar{3}.3402 = 2.19 \times 10^{-3} = 0.00219$.

Powers of Numbers. A number may be squared, cubed, or raised to any power by multiplying the logarithm of the number by the power desired and finding the antilogarithm of the product.

Example 1. Calculate the result of squaring 2 (2^2).

$$\log 2 = 0.3010.$$

Multiply the logarithm of 2 by the power desired, in this case 2.

$$0.3010 \times 2 = 0.6020.$$

The antilogarithm of 0.6020 is 4.00.

Example 2. Find the value of 14^5 (14 taken as a factor 5 times).

$$\log 14 = 1.1461.$$

Multiply the logarithm of 14 by the power desired, in this case 5:

$$1.1461 \times 5 = 5.7305.$$

The antilogarithm is the value sought:

$$\text{Antilog of } 5.7305 = 5.377 \times 10^5 = 537{,}700.$$

In taking powers and roots the characteristic is multiplied by or divided by the power or root along with the mantissa as any number containing a decimal is multiplied or divided.

Roots. The root of a number is found by dividing the logarithm of the number by the root desired. If the square root (second root) is desired, divide the logarithm of the number by 2. If the 4th root is desired, the logarithm must be divided by 4. The antilogarithm of the result is the desired root.

Example 1. Find the cube (3rd) root of 857.

$$\log 857 = 2.9330.$$

Divide by 3 to get the logarithm of the third root:

$$\frac{2.9330}{3} = 0.9777.$$

The antilogarithm of 0.9777 = 9.5.

Example 2. Calculate the 5th root of 3.46×10^{-8}.

$$\log 3.46 \times 10^{-8} = \bar{8}.5391.$$

The entire logarithm, characteristic and mantissa, must be made negative before dividing by five:

$$-8 + .5391 = -7.4609.$$

$$\frac{-7.4609}{5} = -1.4922.$$

There are no negative mantissas, as in -1.4922, in tables. The mantissa must be made positive by making the characteristic one unit more negative and subtracting the negative mantissa. This results in the value $\bar{2}.5078$. The antilog of $\bar{2}.5078 = 3.22 \times 10^{-2} = 0.0322$.

APPENDIX I

Answers to Problems

Chapter 1.

6. HCl — 35.46 g
   HNO_3 — 63.0 g
   H_2SO_4 — 49.04 g
   HNO_3 — 21.0 g as an oxidizer
7. 49.0 g
8. 31.61 g
9. 229.1 ml
10. 0.1047 N
11. Coefficients for each substance in the order it occurs in the equation
    a. 5, 2, 6 → 2, 5, 4
    b. 3, 1, 8 → 1, 3, 12
    c. 2, 1, 4 → 2, 1, 6
    d. 2, 3, 4 → 2, 2
    e. 1, 2 → 1, 1, 1, 1
    f. 2, 1, 4 → 1, 1

Chapter 2.

1. $K_E = \frac{[C][D]^3}{[A][B]^2}$
2. 1×10^{-8} M
   1×10^{-16} K_{sp}
3. 2.93×10^{-2} M
   1×10^{-4} K_{sp}
4. 1.52×10^{-3} moles per liter
   0.7 g per liter
5. 0.0533 g per 100 ml
6. Monochlorotetrammine-cobalt(II)
   Dichlorotetramminecobalt(II)
   Dichlorodiammineplatinum(II
   Trioxalatoferrate(III)
   Hexafluoroaluminate(III)
   Hexacyanoferrate(III)
   Tricyanocuprate(I)
   Hexacyanoferrate(II)
8. 6.67×10^{-12}
9. 18
10. 1.5×10^{-5} M
13. a. 2.7
    b. 12.7
    c. 12.82
14. 2.93
15. 3.95
16. 4.09
17. 2.72
18. 4.15
19. 1.48×10^{-22}
20. 13.5 M
21. 5.84×10^{-20} M
22. 1.58×10^{-15} M
23. Concentration of ion form and molecular form must be equal.

Chapter 3.

4. 0.545 M in H_3O^+ and Cl^-
5. $[Ag^+] = 2.2 \times 10^{-10}$.
   $[Pb^{+2}] = 3.36 \times 10^{-4}$.

Chapter 4.

4. a. $HgCl_2$
   b. HCl
   c. SO_4^{-2}
   d. Any cation of Group II
   e. $SnCl_6^{-2}$
   f. Pb^{+2}
   g. H_2O_2
   h. NH_3
   i. HCl
   j. SO_4^{-2}
   k. HCl
   l. $NH_4C_2H_3O_2$

5. As, Sb, and Sn. As_2S_3 and SnS_2 are yellow, and the orange of Sb_2S_3 may be hidden by greater amounts of As or Sn.
6. a. Cu: green solution, black sulfide. Sb: orange sulfide in arsenic division.
   b. Hg is absent because the black sulfide solid is soluble in 3 M HNO_3.
   c. Pb, Bi, Cd, As, and Sn.
7. a. Cu: no green or blue color in solution; no black precipitate in Step 7; no blue color in Step 13. Hg: no black precipitate in Step 7; sulfides are soluble in Step 9. Pb: no black precipitate in Step 7; no precipitate with sulfate in Step 11. As: no sulfide left insoluble in Step 18. Bi: no dark precipitate in Step 7; no white precipitate insoluble in NH_3 solution in Step 13.
   b. Sb: the orange of Sb_2S_3 usually shows up in Step 17, but it may be hidden if not present in large amounts.
   c. Cd and Sn: cadmium hydroxide dissolves in excess ammonia solution to give a colorless solution; tin sulfide dissolves in concentrated HCl in Step 18.

Chapter 5.

5. a. All elements of the nickel division; their hydroxides are not amphoteric. These are Ni, Co, Mn, and Fe. In addition, since the solution is colorless, all colored ions are absent. These are Ni^{+2}, Co^{+2}, Fe^{+3}, and Cr^{+3} or CrO_4^{-2}.
   b. Al^{+3} and Zn^{+2} are the only possible elements present, since their hydroxides are amphoteric and the ions are colorless.
8. a. The green color of the solution of the sample indicates that Ni^{+2} or Cr^{+3} or both might be present. That NH_3 solution gives a precipitate which is dark green indicates Cr^{+3}, although $Ni(OH)_2$ is green. The peroxide treatment gives a yellow solution like chromate which strongly indicates Cr^{+3} was present. The black precipitate from the $(NH_4)_2S$ treatment might be CoS or NiS or both.
   b. No precipitate with $KClO_3$ indicates the absence of Mn^{+2}. No precipitate with NH_3 indicates the absence of Fe^{+3}. No precipitate with acetic acid, acetate, and KNO_2 indicates the absence of cobalt.
   c. The green color of the solution of the sample, the dark solid remaining after the NaOH, Na_2O_2 treatment, and the absence of a cobalt test all indicate that a positive test for Ni^{+2} is to be expected.

Chapter 7.

7. a. $Co(NO_2)_6^{-3}$ or $Na_3Co(NO_2)_6$
   b. $Co(NO_2)_6^{-3}$
   c. $Zn(UO_2)_3(C_2H_3O_2)_9^-$

Chapter 9.

8. $Zn(OH)_2$. The concentration of Zn^{+2} needed to saturate the solution with $ZnCO_3$ is 1×10^{-10} and that needed to saturate it with $Zn(OH)_2$ is 1×10^{-13}.

11. a. For Cl^-, $[Ag^+] = 2.4 \times 10^{-9}$.
    For PO_4^{-3}, $[Ag^+] = 2.9 \times 10^{-6}$.
    For S^{-2}, $[Ag^+] = 1.2 \times 10^{-24}$.
    For CN^-, $[Ag^+] = 6 \times 10^{-11}$.

    b. In the order 1. Ag_2S, 2. $AgCN$, 3. $AgCl$, and 4. Ag_3PO_4.

Shelf Reagents

ACIDS

Acid	*Mol. Wt.*	*Density*	*%*	*Concentration*	*Method of Preparation*
Acetic	60.05	1.05	99.5	17.5 M	Commercial glacial
Hydrochloric	36.46	1.2	38	12 M	Commercial concentrated
Hydrochloric			19	6 M	One part concentrated to one part water
Hydrochloric			9.5	3 M	One part concentrated to three parts water
Nitric	63	1.4	69	15 M	Commercial concentrated
Nitric			27.5	6 M	400 ml of concentrated added to water and diluted to 1 liter
Nitric			14	3 M	200 ml of concentrated added to water and diluted to 1 liter
Sulfuric	98.1	1.83	98	18 M	Commercial concentrated
Sulfuric			33	6 M	330 ml of concentrated added slowly to water and diluted to 1 liter
Sulfuric			16.3	3 M	165 ml of concentrated added slowly to water and diluted to 1 liter

BASES

Base					
Ammonia	17	0.90	28	15 M	Commercial concentrated ammonium hydroxide
Ammonia			11	6 M	400 ml of concentrated added to water and diluted to 1 liter
Ammonia			5.6	3 M	200 ml of concentrated added to water and diluted to 1 liter
Potassium hydroxide	56.1		34	6 M	340 g solid dissolved in water and diluted to 1 liter
Sodium hydroxide	40		24	6 M	240 g dissolved in water and diluted to 1 liter

SOLUTIONS

Aluminon. Aurin tricarboxylic acid ammonium or sodium salt. 0.1%
1 g dissolved in 1 liter of water

Ammonium acetate. $NH_4C_2H_3O_2$. 3 M
231 g dissolved in water and diluted to 1 liter

Ammonium carbonate. $(NH_4)_2CO_3$. 2 M
192 g $(NH_4)_2CO_3$, 75 ml of concentrated NH_3 diluted to 1 liter with water

Ammonium chloride. NH_4Cl. 4 M
214 g dissolved in water and diluted to 1 liter

Ammonium molybdate. $(NH_4)_2MoO_4$. 0.5 M
90 g of $(NH_4)_2Mo_7O_{24} \cdot 4H_2O$, 35 ml of concentrated NH_3, 240 g of NH_4NO_3 dissolved and diluted to 1 liter

Ammonium nitrate. NH_4NO_3. 1 M
80 g dissolved in water and diluted to 1 liter

Ammonium oxalate. $(NH_4)_2C_2O_4$. 0.3 M
38 g dissolved in water and diluted to 1 liter

Ammonium sulfate. $(NH_4)_2SO_4$. 0.2 M
26 g dissolved in water and diluted to 1 liter

Ammonium sulfide. $(NH_4)_2S$.
One part reagent grade ammonium sulfide plus 2 parts of water

Barium chloride. $BaCl_2$. 0.2 M
49 g of $BaCl_2 \cdot 2H_2O$ dissolved in water and diluted to 1 liter

Barium hydroxide. $Ba(OH)_2$. Saturated
Sufficient $Ba(OH)_2 \cdot 8H_2O$ to saturate; protected from CO_2 of the air

Bromine water. Br_2. Saturated
Bromine shaken with water until saturated

Calcium chloride. $CaCl_2$. 0.3 M
42 g of $CaCl_2 \cdot 2H_2O$ dissolved in water and diluted to 1 liter

Dimethyl glyoxime. $CH_3C(NOH)C(NOH)CH_3$. 1.0%
10 g dissolved in alcohol and diluted to 1 liter with alcohol (95% ethyl)

Disodium hydrogen phosphate. Na_2HPO_4. 0.2 M
28 g dissolved in water and diluted to 1 liter

Ferric chloride. $FeCl_3$. 0.1 M
27 g of $FeCl_3 \cdot 6H_2O$ dissolved in water, with 4 or 5 ml of concentrated HCl added, and diluted to 1 liter with water

Hydrogen peroxide. H_2O_2. 3%

Lead acetate. $Pb(C_2H_3O_2)_2 \cdot 3H_2O$. 0.5 M
190 g dissolved in water and diluted to 1 liter

Magnesia mixture. $MgCl_2$ and NH_4Cl
55 g of $MgCl_2 \cdot 6H_2O$, 140 g of NH_4Cl, and 131 ml of concentrated NH_3 diluted to 1 liter with water

SOLUTIONS (Cont.)

Magnesium reagent. 0.01%

0.1 g of *p*-nitrobenzeneazoresorcinol in 1 liter of water made 0.025 M in NaOH by adding 1 gram of solid NaOH to 1 liter of water

Mercuric chloride. $HgCl_2$. 0.2 M

54 g dissolved in water and diluted to 1 liter

Potassium acetate. $KC_2H_3O_2$. 2 M

196 g dissolved in water and diluted to 1 liter

Potassium chromate. K_2CrO_4. 0.5 M

97 g dissolved in water and diluted to 1 liter

Potassium cyanide. KCN. 0.2 M

13 g dissolved in water and diluted to 1 liter

Potassium ferrocyanide. Potassium hexacyanoferrate(II).

$$K_4Fe(CN)_6 \cdot 3H_2O. \quad 0.2\ M$$

84 g dissolved in water and diluted to 1 liter

Potassium iodide. KI. 0.5 M

83 g dissolved in water and diluted to 1 liter

Potassium nitrite. KNO_2. 0.2 M

17 g dissolved in water and diluted to 1 liter

Potassium nitrite. KNO_2. 6 M

510 g dissolved in water and diluted to 1 liter

Potassium permanganate. $KMnO_4$. 0.1 M

15 g dissolved in water and diluted to 1 liter

Potassium thiocyanate. KSCN. 0.5 M

49 g dissolved in water and diluted to 1 liter

Silver nitrate. $AgNO_3$. 0.5 M

85 g dissolved in water and diluted to 1 liter

Silver sulfate. Ag_2SO_4. Saturated

About 8 g dissolved in 1 liter of water

Sodium carbonate. Na_2CO_3. Saturated

About 220 g dissolved in 1 liter of water

Sodium cobaltinitrite. Sodium hexanitrocobaltate(III). 10%

100 g dissolved in water and diluted to 1 liter

Stannous chloride. $SnCl_2 \cdot 2H_2O$. 0.2 M

45 g $SnCl_2 \cdot 2H_2O$. Add 250 ml concentrated HCl and enough water to dissolve. Cool and dilute to 1 liter.

Thioacetamide. CH_3CSNH_2. 10%

100 g dissolved in water and diluted to 1 liter. Keep cool.

Zinc uranyl acetate. Sodium reagent.

10 g of $UO_2(C_2H_3O_2)_2 \cdot 2H_2O$ plus 5 ml of 5 M $HC_2H_3O_2$ plus H_2O to make about 50 ml. Stir thoroughly, add 30 g of $Zn(C_2H_3O_2)_2 \cdot 2H_2O$, and dilute to 1 liter.

Liquids

Amyl alcohol. $C_5H_{12}OH$
Carbon tetrachloride. CCl_4
Ether, diethyl. $(C_2H_5)_2O$
Ethyl alcohol. C_2H_5OH
Methyl alcohol. CH_3OH

Solids

Ammonium sulfate. $(NH_4)_2SO_4$
Ammonium thiocyanate. NH_4NCS
Borax. $Na_2B_4O_7 \cdot 10H_2O$
Ferrous sulfate. Iron(II) sulfate. $FeSO_4 \cdot 7H_2O$
Iron wire or brads. Fe
Oxalic acid. $H_2C_2O_4$
Potassium chlorate. $KClO_3$
Sodium bismuthate. $NaBiO_3$
Sodium carbonate. Na_2CO_3
Sodium dithionate. $Na_2S_2O_4$
Sodium fluoride. NaF
Sodium hydroxide. NaOH
Tin (mossy). Sn
Zinc (granules). Zn

APPENDIX III

Instability Constants

Equilibrium	K_{Inst}
$Ag(NH_3)_2^+ \rightleftharpoons Ag^+ + 2NH_3$	6×10^{-8}
$Cu(NH_3)_4^{+2} \rightleftharpoons Cu^{+2} + 4NH_3$	5×10^{-14}
$Cd(NH_3)_4^{+2} \rightleftharpoons Cd^{+2} + 4NH_3$	2.4×10^{-7}
$Cu(CN)_3^{-2} \rightleftharpoons Cu^+ + 3CN^-$	5×10^{-28}
$Cd(CN)_4^{-2} \rightleftharpoons Cd^{+2} + 4CN^-$	1.4×10^{-17}
$Al(OH)_4^- \rightleftharpoons \underline{Al(OH)_3} + OH^-$	2.5×10^{-2}
$FeSCN^{+2} \rightleftharpoons Fe^{+3} + SCN^-$	7.3×10^{-3}

APPENDIX IV

Some Acid-Base Indicators

Indicator	Acid Color	Alkaline Color	pH Range of Change
Picric acid	Colorless	Yellow	0.1– 0.8
Thymol blue	Red	Yellow	1.2– 2.8
2,6-Dinitrophenol	Colorless	Yellow	2.0– 4.0
Congo red	Blue	Red	3.0– 5.0

Indicator	Acid Color	Alkaline Color	pH Range of Change
Methyl orange	Red	Yellow	3.2– 4.4
Bromcresol green	Yellow	Blue	3.8– 5.4
Methyl red	Red	Yellow	4.2– 6.2
Methyl purple	Purple	Green	4.8– 5.4
Bromthymol blue	Yellow	Blue	6.0– 7.6
Litmus	Red	Blue	4.5– 8.3
Phenol red	Yellow	Red	6.6– 8.4
Phenolphthalein	Colorless	Red	8.3–10.0
Thymolphthalein	Colorless	Blue	9.3–10.5
Alizarin yellow	Yellow	Violet	10.1–12.0
2,4,6-Trinitrotoluene	Colorless	Orange	11.5–13.0

APPENDIX V

Ionization Constants of Acids and Bases

Acid		Equilibrium Reaction		K_A
Acetic		$HC_2H_3O_2 + H_2O \rightleftarrows H_3O^+ + C_2H_3O_2^-$		1.8×10^{-5}
Arsenic	(1)	$H_3AsO_4 + H_2O \rightleftarrows H_3O^+ + H_2AsO_4^-$	K_1	2.8×10^{-4}
	(2)	$H_2AsO_4^- + H_2O \rightleftarrows H_3O^+ + HAsO_4^{-2}$	K_2	7×10^{-8}
	(3)	$HAsO_4^{-2} + H_2O \rightleftarrows H_3O^+ + AsO_4^{-3}$	K_3	1×10^{-13}
Carbonic	(1)	$H_2CO_3 + H_2O \rightleftarrows H_3O^+ + HCO_3^-$	K_1	3.1×10^{-7}
	(2)	$HCO_3^- + H_2O \rightleftarrows H_3O^+ + CO_3^{-2}$	K_2	4.2×10^{-11}
Hydrocyanic		$HCN + H_2O \rightleftarrows H_3O^+ + CN^-$		1×10^{-9}
Hydrofluoric		$HF + H_2O \rightleftarrows H_3O^+ + F^-$		1.7×10^{-5}
Hydrogen sulfide	(1)	$H_2S + H_2O \rightleftarrows H_3O^+ + HS^-$	K_1	5.9×10^{-8}
	(2)	$HS^- + H_2O \rightleftarrows H_3O^+ + S^{-2}$	K_2	1×10^{-15}
Lactic		$HC_3H_5O_3 + H_2O \rightleftarrows H_3O^+ + C_3H_5O_3^-$		1.4×10^{-4}
Oxalic	(1)	$H_2C_2O_4 + H_2O \rightleftarrows H_3O^+ + HC_2O_4^-$	K_1	5×10^{-2}
	(2)	$HC_2O_4^- + H_2O \rightleftarrows H_3O^+ + C_2O_4^{-2}$	K_2	3×10^{-5}
Phosphoric	(1)	$H_3PO_4 + H_2O \rightleftarrows H_3O^+ + H_2PO_4^-$	K_1	7.5×10^{-3}
	(2)	$H_2PO_4^- + H_2O \rightleftarrows H_3O^+ + HPO_4^{-2}$	K_2	6.2×10^{-8}
	(3)	$HPO_4^{-2} + H_2O \rightleftarrows H_3O^+ + PO_4^{-3}$	K_3	2×10^{-13}
Phthalic	(1)	$H_2C_8H_4O_4 + H_2O \rightleftarrows H_3O^+ + HC_8H_4O_4^-$	K_1	1.29×10^{-3}
	(2)	$HC_8H_4O_4^- + H_2O \rightleftarrows H_3O^+ + C_8H_4O_4^{-2}$	K_2	3.8×10^{-6}

Base	Equilibrium Reaction	K_B
Ammonia (ammonium hydroxide)	$NH_3 + H_2O \rightleftarrows NH_4^+ + OH^-$	1.8×10^{-5}
Aniline	$C_6H_5NH_2 + H_2O \rightleftarrows C_6H_5NH_3^+ + OH^-$	2.7×10^{-10}
Diethylamine	$(C_2H_5)_2NH + H_2O \rightleftarrows (C_2H_5)_2NH_2^+ + OH^-$	8.92×10^{-4}
Hydrazine	$N_2H_4 + H_2O \rightleftarrows N_2H_5^+ + OH^-$	3×10^{-6}
Triethylamine	$(C_2H_5)_3N + H_2O \rightleftarrows (C_2H_5)_3NH^+ + OH^-$	6.5×10^{-4}

Solubility Product Constants

Substance	Formula	K_{sp}
Aluminum hydroxide	$Al(OH)_3$	4×10^{-33}
Antimony sulfide	Sb_2S_3	3×10^{-53}
Arsenic sulfide	As_2S_3	1.1×10^{-33}
Barium		
carbonate	$BaCO_3$	5×10^{-9}
chromate	$BaCrO_4$	1.2×10^{-10}
fluoride	BaF_2	2×10^{-5}
oxalate	BaC_2O_4	1.6×10^{-8}
phosphate	$Ba_3(PO_4)_2$	4×10^{-39}
sulfate	$BaSO_4$	1.1×10^{-10}
sulfite	$BaSO_3$	1×10^{-9}
Bismuth sulfide	Bi_2S_3	7×10^{-97}
Cadmium		
carbonate	$CdCO_3$	5×10^{-12}
sulfide	CdS	1×10^{-26}
Calcium		
carbonate	$CaCO_3$	5×10^{-9}
fluoride	CaF_2	4×10^{-11}
oxalate	CaC_2O_4	2.6×10^{-9}
phosphate	$Ca_3(PO_4)_2$	1×10^{-32}
sulfate	$CaSO_4$	6×10^{-5}
Chromium hydroxide	$Cr(OH)_3$	7×10^{-31}
Cobalt		
hydroxide	$Co(OH)_2$	2.5×10^{-16}
sulfide	CoS	1×10^{-21}
Copper		
chloride	$CuCl$	4×10^{-7}
sulfide	Cu_2S	2×10^{-48}
Copper		
carbonate	$CuCO_3$	2.5×10^{-5}
sulfide	CuS	9×10^{-36}
Ferric		
hydroxide	$Fe(OH)_3$	6×10^{-38}
sulfide	Fe_2S_3	1×10^{-88}
Ferrous		
carbonate	$FeCO_3$	2×10^{-11}
sulfide	FeS	5×10^{-18}
Lead		
carbonate	$PbCO_3$	1.2×10^{-13}
chloride	$PbCl_2$	1×10^{-4}
chromate	$PbCrO_4$	2×10^{-16}

fluoride	PbF_2	4×10^{-8}
hydroxide	$Pb(OH)_2$	2×10^{-15}
iodide	PbI_2	1.4×10^{-8}
sulfate	$PbSO_4$	1.1×10^{-8}
sulfide	PbS	8×10^{-28}
Magnesium		
ammonium phosphate	$MgNH_4PO_4$	2.5×10^{-13}
carbonate	$MgCO_3$	4×10^{-5}
hydroxide	$Mg(OH)_2$	1×10^{-11}
oxalate	MgC_2O_4	8.7×10^{-5}
Manganous		
carbonate	$MnCO_3$	9×10^{-11}
hydroxide	$Mn(OH)_2$	1×10^{-13}
sulfide	MnS	5×10^{-15}
Mercuric sulfide	HgS	1×10^{-54}
Mercurous		
chloride	Hg_2Cl_2	1.3×10^{-18}
sulfide	Hg_2S	6×10^{-44}
Nickel		
carbonate	$NiCO_3$	1.4×10^{-7}
sulfide	NiS	2×10^{-21}
Silver		
arsenate	Ag_3AsO_4	1×10^{-22}
bromide	$AgBr$	7.7×10^{-13}
chloride	$AgCl$	1.2×10^{-10}
chromate	Ag_2CrO_4	1.2×10^{-12}
cyanide	$AgCN$	3×10^{-12}
iodide	AgI	1×10^{-16}
phosphate	Ag_3PO_4	1.2×10^{-18}
sulfate	Ag_2SO_4	1.2×10^{-5}
sulfide	Ag_2S	7×10^{-50}
thiocyanate	$AgSCN$	2×10^{-12}
Stannous		
hydroxide	$Sn(OH)_2$	8×10^{-27}
sulfide	SnS	1×10^{-25}
Strontium		
carbonate	$SrCO_3$	8×10^{-10}
fluoride	SrF_2	9×10^{-10}
phosphate	$Sr_3(PO_4)_2$	1×10^{-31}
sulfate	$SrSO_4$	5×10^{-7}
Zinc		
carbonate	$ZnCO_3$	2×10^{-10}
hydroxide	$Zn(OH)_2$	4×10^{-17}
sulfide	ZnS	1×10^{-21}

APPENDIX VII

International Atomic Weights (1961)

Element	Symbol	Atomic No.	Atomic Weight	Element	Symbol	Atomic No.	Atomic Weight
Actinium	Ac	89		Mercury	Hg	80	200.59
Aluminum	Al	13	26.9815	Molybdenum	Mo	42	95.94
Americium	Am	95		Neodymium	Nd	60	144.24
Antimony	Sb	51	121.75	Neon	Ne	10	20.183
Argon	Ar	18	39.948	Neptunium	Np	93	
Arsenic	As	33	74.9216	Nickel	Ni	28	58.71
Astatine	At	85		Niobium	Nb	41	92.906
Barium	Ba	56	137.34	Nitrogen	N	7	14.0067
Berkelium	Bk	97		Nobelium	No	102	
Beryllium	Be	4	9.0122	Osmium	Os	76	190.2
Bismuth	Bi	83	208.980	Oxygen	O	8	15.9994[a]
Boron	B	5	10.811[a]	Palladium	Pd	46	106.4
Bromine	Br	35	79.909[b]	Phosphorus	P	15	30.9738
Cadmium	Cd	48	112.40	Platinum	Pt	78	195.09
Calcium	Ca	20	40.08	Plutonium	Pu	94	
Californium	Cf	98		Polonium	Po	84	
Carbon	C	6	12.01115[a]	Potassium	K	19	39.102
Cerium	Ce	58	140.12	Praseodymium	Pr	59	140.907
Cesium	Cs	55	132.905	Promethium	Pm	61	
Chlorine	Cl	17	35.453[b]	Protactinium	Pa	91	
Chromium	Cr	24	51.996[b]	Radium	Ra	88	
Cobalt	Co	27	58.9332	Radon	Rn	86	
Copper	Cu	29	63.54	Rhenium	Re	75	186.2
Curium	Cm	96		Rhodium	Rh	45	102.905
Dysprosium	Dy	66	162.50	Rubidium	Rb	37	85.47
Einsteinium	Es	99		Ruthenium	Ru	44	101.07
Erbium	Er	68	167.26	Samarium	Sm	62	150.35
Europium	Eu	63	151.96	Scandium	Sc	21	44.956
Fermium	Fm	100		Selenium	Se	34	78.96
Fluorine	F	9	18.9984	Silicon	Si	14	28.086[a]
Francium	Fr	87		Silver	Ag	47	107.870[b]
Gadolinium	Gd	64	157.25	Sodium	Na	11	22.9898
Gallium	Ga	31	69.72	Strontium	Sr	38	87.62
Germanium	Ge	32	72.59	Sulfur	S	16	32.064[a]
Gold	Au	79	196.967	Tantalum	Ta	73	180.948
Hafnium	Hf	72	178.49	Technetium	Tc	43	
Helium	He	2	4.0026	Tellurium	Te	52	127.60
Holmium	Ho	67	164.930	Terbium	Tb	65	158.924
Hydrogen	H	1	1.00797[a]	Thallium	Tl	81	204.37
Indium	In	49	114.82	Thorium	Th	90	232.038
Iodine	I	53	126.9044	Thulium	Tm	69	168.934
Iridium	Ir	77	192.2	Tin	Sn	50	118.69
Iron	Fe	26	55.847[b]	Titanium	Ti	22	47.90
Krypton	Kr	36	83.80	Tungsten	W	74	183.85
Lanthanum	La	57	138.91	Uranium	U	92	238.03
Lead	Pb	82	207.19	Vanadium	V	23	50.942
Lithium	Li	3	6.939	Xenon	Xe	54	131.30
Lutetium	Lu	71	174.97	Ytterbium	Yb	70	173.04
Magnesium	Mg	12	24.312	Yttrium	Y	39	88.905
Manganese	Mn	25	54.9380	Zinc	Zn	30	65.37
Mendelevium	Md	101		Zirconium	Zr	40	91.22

[a] The atomic weight varies because of natural variations in the isotopic composition of the element. The observed ranges are boron, ±0.003; carbon, ±0.00005; hydrogen, ±0.00001; oxygen, ±0.0001; silicon, ±0.001; sulfur, ±0.003.

[b] The atomic weight is believed to have an experimental uncertainty of the following magnitude; bromine, ±0.002; chlorine, ±0.001; chromium, ±0.001; iron, ±0.003; silver, ±0.003. For other elements the last digit given is believed to be reliable to ±0.5.

APPENDIX VIII

Table of Logarithms

N	0	1	2	3	4	5	6	7	8	9	Proportional Parts 1	2	3	4	5
10	0000	0043	0086	0128	0170	0212	0253	0294	0334	0374	4	8	12	17	21
11	0414	0453	0492	0531	0569	0607	0645	0682	0719	0755	4	8	11	15	19
12	0792	0828	0864	0899	0934	0969	1004	1038	1072	1106	3	7	10	14	17
13	1139	1173	1206	1239	1271	1303	1335	1367	1399	1430	3	6	10	13	16
14	1461	1492	1523	1553	1584	1614	1644	1673	1703	1732	3	6	9	12	15
15	1761	1790	1818	1847	1875	1903	1931	1959	1987	2014	3	6	8	11	14
16	2041	2068	2095	2122	2148	2175	2201	2227	2253	2279	3	5	8	11	13
17	2304	2330	2355	2380	2405	2430	2455	2480	2504	2529	2	5	7	10	12
18	2553	2577	2601	2625	2648	2672	2695	2718	2742	2765	2	5	7	9	12
19	2788	2810	2833	2856	2878	2900	2923	2945	2967	2989	2	4	7	9	11
20	3010	3032	3054	3075	3096	3118	3139	3160	3181	3201	2	4	6	8	11
21	3222	3243	3263	3284	3304	3324	3345	3365	3385	3404	2	4	6	8	10
22	3424	3444	3464	3483	3502	3522	3541	3560	3579	3598	2	4	6	8	10
23	3617	3636	3655	3674	3692	3711	3729	3747	3766	3784	2	4	5	7	9
24	3802	3820	3838	3856	3874	3892	3909	3927	3945	3962	2	4	5	7	9
25	3979	3997	4014	4031	4048	4065	4082	4099	4116	4133	2	3	5	7	9
26	4150	4166	4183	4200	4216	4232	4249	4265	4281	4298	2	3	5	7	8
27	4314	4330	4346	4362	4378	4393	4409	4425	4440	4456	2	3	5	6	8
28	4472	4487	4502	4518	4533	4548	4564	4579	4594	4609	2	3	5	6	8
29	4624	4639	4654	4669	4683	4698	4713	4728	4742	4757	1	3	4	6	7
30	4771	4786	4800	4814	4829	4843	4857	4871	4886	4900	1	3	4	6	7
31	4914	4928	4942	4955	4969	4983	4997	5011	5024	5038	1	3	4	6	7
32	5051	5065	5079	5092	5105	5119	5132	5145	5159	5172	1	3	4	5	7
33	5185	5198	5211	5224	5237	5250	5263	5276	5289	5302	1	3	4	5	6
34	5315	5328	5340	5353	5366	5378	5391	5403	5416	5428	1	3	4	5	6
35	5441	5453	5465	5478	5490	5502	5514	5527	5539	5551	1	2	4	5	6
36	5563	5575	5587	5599	5611	5623	5635	5647	5658	5670	1	2	4	5	6
37	5682	5694	5705	5717	5729	5740	5752	5763	5775	5786	1	2	3	5	6
38	5798	5809	5821	5832	5843	5855	5866	5877	5888	5899	1	2	3	5	6
39	5911	5922	5933	5944	5955	5966	5977	5988	5999	6010	1	2	3	4	6
40	6021	6031	6042	6053	6064	6075	6085	6096	6107	6117	1	2	3	4	5
41	6128	6138	6149	6160	6170	6180	6191	6201	6212	6222	1	2	3	4	5
42	6232	6243	6253	6263	6274	6284	6294	6304	6314	6325	1	2	3	4	5
43	6335	6345	6355	6365	6375	6385	6395	6405	6415	6425	1	2	3	4	5
44	6435	6444	6454	6464	6474	6484	6493	6503	6513	6522	1	2	3	4	5
45	6532	6542	6551	6561	6571	6580	6590	6599	6609	6618	1	2	3	4	5
46	6628	6637	6646	6656	6665	6675	6684	6693	6702	6712	1	2	3	4	5
47	6721	6730	6739	6749	6758	6767	6776	6785	6794	6803	1	2	3	4	5
48	6812	6821	6830	6839	6848	6857	6866	6875	6884	6893	1	2	3	4	4
49	6902	6911	6920	6928	6937	6946	6955	6964	6972	6981	1	2	3	4	4
50	6990	6998	7007	7016	7024	7033	7042	7050	7059	7067	1	2	3	3	4
51	7076	7084	7093	7101	7110	7118	7126	7135	7143	7152	1	2	3	3	4
52	7160	7168	7177	7185	7193	7202	7210	7218	7226	7235	1	2	2	3	4
53	7243	7251	7259	7267	7275	7284	7292	7300	7308	7316	1	2	2	3	4
54	7324	7332	7340	7348	7356	7364	7372	7380	7388	7396	1	2	2	3	4
N	0	1	2	3	4	5	6	7	8	9	1	2	3	4	5

Table of Logarithms

(continued)

N	0	1	2	3	4	5	6	7	8	9	Proportional Parts 1	2	3	4	5
55	7404	7412	7419	7427	7435	7443	7451	7459	7466	7474	1	2	2	3	4
56	7482	7490	7497	7505	7513	7520	7528	7536	7543	7551	1	2	2	3	4
57	7559	7566	7574	7582	7589	7597	7604	7612	7619	7627	1	2	2	3	4
58	7634	7642	7649	7657	7664	7672	7679	7686	7694	7701	1	1	2	3	4
59	7709	7716	7723	7731	7738	7745	7752	7760	7767	7774	1	1	2	3	4
60	7782	7789	7796	7803	7810	7818	7825	7832	7839	7846	1	1	2	3	4
61	7853	7860	7868	7875	7882	7889	7896	7903	7910	7917	1	1	2	3	4
62	7924	7931	7938	7945	7952	7959	7966	7973	7980	7987	1	1	2	3	3
63	7993	8000	8007	8014	8021	8028	8035	8041	8048	8055	1	1	2	3	3
64	8062	8069	8075	8082	8089	8096	8102	8109	8116	8122	1	1	2	3	3
65	8129	8136	8142	8149	8156	8162	8169	8176	8182	8189	1	1	2	3	3
66	8195	8202	8209	8215	8222	8228	8235	8241	8248	8254	1	1	2	3	3
67	8261	8267	8274	8280	8287	8293	8299	8306	8312	8319	1	1	2	3	3
68	8325	8331	8338	8344	8351	8357	8363	8370	8376	8382	1	1	2	3	3
69	8388	8395	8401	8407	8414	8420	8426	8432	8439	8445	1	1	2	3	3
70	8451	8457	8463	8470	8476	8482	8488	8494	8500	8506	1	1	2	2	3
71	8513	8519	8525	8531	8537	8543	8549	8555	8561	8567	1	1	2	2	3
72	8573	8579	8585	8591	8597	8603	8609	8615	8621	8627	1	1	2	2	3
73	8633	8639	8645	8651	8657	8663	8669	8675	8681	8686	1	1	2	2	3
74	8692	8698	8704	8710	8716	8722	8727	8733	8739	8745	1	1	2	2	3
75	8751	8756	8762	8768	8774	8779	8785	8791	8797	8802	1	1	2	2	3
76	8808	8814	8820	8825	8831	8837	8842	8848	8854	8859	1	1	2	2	3
77	8865	8871	8876	8882	8887	8893	8899	8904	8910	8915	1	1	2	2	3
78	8921	8927	8932	8938	8943	8949	8954	8960	8965	8971	1	1	2	2	3
79	8976	8982	8987	8993	8998	9004	9009	9015	9020	9025	1	1	2	2	3
80	9031	9036	9042	9047	9053	9058	9063	9069	9074	9079	1	1	2	2	3
81	9085	9090	9096	9101	9106	9112	9117	9122	9128	9133	1	1	2	2	3
82	9138	9143	9149	9154	9159	9165	9170	9175	9180	9186	1	1	2	2	3
83	9191	9196	9201	9206	9212	9217	9222	9227	9232	9238	1	1	2	2	3
84	9243	9248	9253	9258	9263	9269	9274	9279	9284	9289	1	1	2	2	3
85	9294	9299	9304	9309	9315	9320	9325	9330	9335	9340	1	1	2	2	3
86	9345	9350	9355	9360	9365	9370	9375	9380	9385	9390	1	1	2	2	3
87	9395	9400	9405	9410	9415	9420	9425	9430	9435	9440	0	1	1	2	2
88	9445	9450	9455	9460	9465	9469	9474	9479	9484	9489	0	1	1	2	2
89	9494	9499	9504	9509	9513	9518	9523	9528	9533	9538	0	1	1	2	2
90	9542	9547	9552	9557	9562	9566	9571	9576	9581	9586	0	1	1	2	2
91	9590	9595	9600	9605	9609	9614	9619	9624	9628	9633	0	1	1	2	2
92	9638	9643	9647	9652	9657	9661	9666	9671	9675	9680	0	1	1	2	2
93	9685	9689	9694	9699	9703	9708	9713	9717	9722	9727	0	1	1	2	2
94	9731	9736	9741	9745	9750	9754	9759	9763	9768	9773	0	1	1	2	2
95	9777	9782	9786	9791	9795	9800	9805	9809	9814	9818	0	1	1	2	2
96	9823	9827	9832	9836	9841	9845	9850	9854	9859	9863	0	1	1	2	2
97	9868	9872	9877	9881	9886	9890	9894	9899	9903	9908	0	1	1	2	2
98	9912	9917	9921	9926	9930	9934	9939	9943	9948	9952	0	1	1	2	2
99	9956	9961	9965	9969	9974	9978	9983	9987	9991	9996	0	1	1	2	2
N	0	1	2	3	4	5	6	7	8	9	1	2	3	4	5

Index

Index